ELECTRICAL
EDUCATION GUIDE

WRITTEN, DESIGNED, AND EDITED BY:

A.M. CAGNOLA

TTI PUBLISHING LLC.

The Texas Trade Institute was established in 2015 with the sole intent of educating the future leaders of the electrical industry. While providing electrical education programs through numerous school systems, the Texas Trade Institute continues to be at the forefront of integrating new technology into interactive and practical education. The strong family ties of our founders, their passion for education, and over half a century of electrical experience is revealed everyday through our colleagues and students.

This work is a reflection of the national standards, codes, and practices of the electrical field. All the procedures and information found in this publication are up to date with the latest National Electric Code standards.

NFPA 70®, *National Electric Code* and *NEC®* are registered trademarks of the National Fire Protection Association, Quincy, MA.

ISBN: 0-9996369-0-9

We would like to extend our sincerest thanks to **Patrick V. Cagnola** for his contribution to this publication. This work would not have been possible without his knowledge as a master electrician for over 40 years. Even after his retirement as an electrical superintendent, he continues to educate and lead new generations of electricians through training programs. To date, he has started the careers of over 1000 electricians.

Chapter Contents

Introduction

The demand for electricity in the United States continues to grow at an exponential rate. It is important that this expansion is supported by educated and well trained skilled labor. The United States Bureau of Labor Statistics projected that there will be a 14% employment growth for electricians by 2024, which is significantly higher than the average professional occupation growth rate. In order to provide the workforce needed to meet the economy's needs, it is crucial to use efficient and detailed training materials to properly educate the future leaders of the electrical industry.

This publication is meant as a guide in your journey to becoming a fully trained electrician. It is best used as a guideline for proper industry practices, and can be used in conjunction with an apprenticeship electrical education program. The information found within this text should provide ample information needed for education, training, and testing.

The information referenced and used to formulate this text is based on the minimum standards, regulations, and codes put forth by the **National Electrical Code (*NEC®*)**. It is utilized as the primary source for proper industry techniques and safety guidelines and is provided by the **National Fire Protection Association (NFPA®)**. Residential, commercial, and industrial electrical installation guidelines and standards are in accordance with this national code, as well as local building codes to ensure the safety of both the workers and the public.

Local codes may override national codes as long as the required local codes meet or exceed that of the requirements put forth by the national code, never less. Below are nationally recognized organizations for the safety analysis of electrical devices and tools. You should familiarize yourself with them for future reference.

(UL) Underwriters Laboratories

Founded in 1894, UL was established in order to research and test standards for workplace safety and compliance, as well as the protection of security and brand reputation. UL provides minimum standards to meet the quality and safety of todays electrical industry. When working with tools or materials, look for the UL stamp of approval to ensure that these items are meeting the required standards. If certain products in the electrical field do not carry the UL seal of approval, then they may be deemed unsafe for the workplace and prohibited from use by the electrical inspector.

(OSHA) Occupational Safety and Health Association

With the passing of the Occupational Safety and Health Act of 1970, the establishment of OSHA was created to "assure safe and healthful working conditions for working men and women". OSHA covers almost all private and public sectors of business under its federal authority. While working, you should adhere to OSHA guidelines to ensure safe workplace practices. OSHA representatives also hold the authority to temporarily stop work on jobsites if violations of OSHA code have taken place. It is your responsibility as an electrician to always carry out safe practices and adhere to OSHA guidelines.

(CSA) Canadian Standards Association

CSA Group is a global organization dedicated to safety, social good and sustainability. CSA Group is one of the largest standards development organizations in North America - conducting research and developing standards for a broad range of technologies and functional areas. CSA Group is also a global provider of testing, inspection, and certification services for products in many market sectors - and a leader in safety and environmental certification for Canada and the US.

(ARL) Applied Research Laboratories

ARL was established in 1945 to conduct research on improving national security through acoustics, electromagnetics, and information science. It is one of the oldest research facilities still associated with its university (University of Texas). The data they provide is often used when rule changes and local codes are being discussed.

These two symbols represent the designation stamped on equipment and tools for approved use. Th left symbol is provided by Underwriters Laboratory (UL), and right symbol is provided by the Canadian Standards Association (CSA).

Electrical Overview

1

All electrical systems have the potential to cause serious injury or death. It is imperative to have a working knowledge of electrical behavior as well as the tools and equipment to control it. Having this knowledge will allow you to properly provide safe working conditions for personnel, and will ensure safe operation of equipment.

Remember, you must be a qualified electrician to work with an electrical circuit. **Do not attempt** to test any electrical circuit or device without understanding the fundamentals of electricity and testing procedures. Always consult your supervisor if you are unsure of certain procedures or guidelines.

Throughout this section, you will gain an understanding of how electrical power is created. In addition, you will learn important electrical symbols you will use on a daily basis along with various wiring plans. These symbols are the set standard for electrical design and will be referenced often within this work.

The remainder of this chapter will cover safety and equipment. Safety is an extremely important part of your ability to carry out your work properly. It is important that you understand the steps you need to take to comply with safety standards. This includes jobsite practices, protective gear requirements, as well as chain of command protocol.

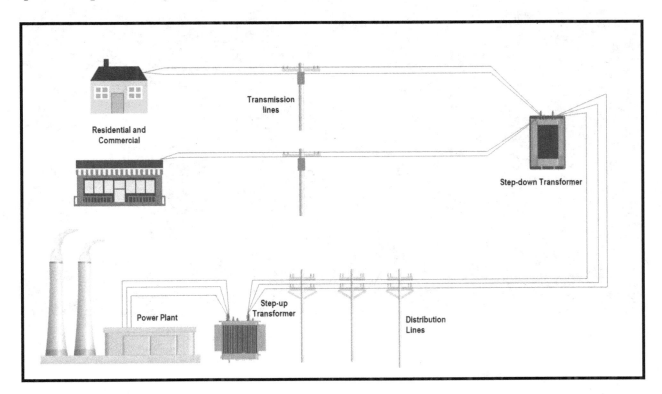

1.1 Electrical Generation

Electrical energy is produced by using natural and renewable resources such as coal, oil, natural gas, wind, hydroelectric, solar, geothermal, and nuclear power. Within the United States, we use nearly 40% of our generated energy to provide electrical power to the United States grid. It is important that you not only understand how these sources produce electrical energy, but also the impact they will have throughout your electrical career.

A. **Solar energy** uses the radiant heat energy generated by the sun and then converts it to electrical energy. Solar energy is a clean and renewable resource, but is not as efficient as coal and oil. Solar energy today accounts for less than 1% of the United States energy consumption. However, is it by far the fastest growing electrical industry. In 2016, nearly 40% of all new electrical installations were solar energy systems. In addition, there are over 9,000 companies that work with solar energy. By 2022, it is predicted that almost 5% of the electric power we receive will be powered by solar energy, an increase of nearly 500%.

B. **Wind power** makes electrical energy by using wind to turn blades on a large wind turbine which in turn spins a generator which produces electrical power. In the last 10 years, more than 143 billion dollars has been invested into the wind power industry. The downside is that it requires areas which have high winds year round to be worth the cost.

C. **Coal, oil, and natural gas** plants account for the largest production of the United States electrical power. These resources are burned in furnaces that heat water and produce steam to operate turbines which generate electricity. It is the industry you will most likely be working with considering it accounts for nearly 80% of the electrical power in the United States. This is the oldest electrical power resource and continues to be the most efficient today. However, as a fossil fuel, there is a limited supply. More than likely there will be a reduction in the use of fossil fuels over the next 100 years.

D. **Hydro power** uses the natural power of large lakes and rivers to turn turbines to generate electricity. This is accomplished by building dams near large water bodies. This can only be done in certain areas where there is enough water flow to create useful energy.

Nuclear power is also an important energy source for our future. Nuclear power is produced through nuclear fission. This occurs when atoms are split apart releasing energy, similar to the way atom bombs work! Most nuclear plants use uranium atoms to produce their energy.

Nuclear energy is often frowned upon because of major accidents that have occurred, such as the nuclear disaster in 2011 in Japan or Chernobyl in 1986. It is the second largest producer of electricity in the United States at nearly 20%. Some countries, such as France, get a majority of their energy from nuclear power, almost 75%!

1.2 Electrical Terms

Alternating Current (AC) - Electrical power in which the flow of electricity switches its direction of flow at intervals in a conductor.

Ammeter - A tool used to measure electrical power in amperes. They are connected in series with the circuit you wish to test.

Ampere - Is a unit of measure for the flow of electrical current.

Arc Fault Circuit Interrupter (AFCI) - A piece of equipment that can protect from arc faults by detection and then de-energization of the circuit.

Cable - When multiple wires or conductors are in a single group and are protected by a covering usually made of rubber.

Capacitor - A device used to store electrical energy to filter out voltage spikes.

Circuit - A certain path which is used to control the flow of electrical power.

Circuit Breaker - A safety device that protects against overcurrent to the circuit.

Conductor - Is a material such as copper or aluminum which is used to move electrical charges.

Current - Movement of electricity along a conductor.

Direct Current (DC) - Electrical power in which the flow of electricity moves nonstop in the same direction along the conductor from a high point to a low point.

Electricity - The flow of electrons from atom to atom in a conductor.

Fuse - A replaceable safety device for an electrical circuit. In the case of an overload, the wire will melt, which will break the circuit.

Generator - A device used to convert mechanical energy into electrical energy by using fuel.

Ground - Occurs when a part of the wire circuit touches a metallic part of the machine frame.

Grounding Conductor - The wire that transmits electrical power to the earth when there is a short circuit.

Grounding Circuit Conductor - Normally a white wire which is used to return the current under zero pressure from the load to the power source.

Ground Fault Circuit Interrupter (GFCI) - A device used to protect against shock from an unexpected short circuit.

Hot Conductor - A wire that carries electrical current to the load, usually a black or red wire.

Insulator - A material or substance used to resist the flow of electricity. Usually protective equipment has some type of insulator material.

Load - Electrical devices which are connected to the ungrounded conductor.

Milliampere - Is 1/1,000 of an ampere.

Motor - Converts electrical energy into mechanical energy.

Multimeter - A device used to read ohms (resistance), voltage (force), or amperes (current) of a circuit.

Neutral Conductor - A conductor or wire that carries the ampere imbalance between the ungrounded conductors.

OHM - Unit of measure for the resistance of flow to an electrical current.

Outlet - Where the current flowing through the circuit is used for receptacles and powered devices.

Pigtail (jumper) - Short wire or conductor used to connect two or more wires to a single screw terminal on a receptacle.

Receptacle - A piece of equipment used to allow access to a circuit by plug-in.

Resistance - An opposing force placed on the circuit to reduce the flow of electrical current.

Service Entrance Panel (SEP) - A panelboard which is located in the area which receives electrical power from the electrical supplier. It is the main source for circuits run throughout the building.

Service Meter - Device for measurement of the total electrical current used over a period of time by a structure.

Short - Occurs when a part of the circuit touches another part of the same circuit, which sends electrical current in a path not intended.

Subpanel - Another panelboard that is fed from the service entrance panel to deliver more branch circuits.

Switch - Device that can control the flow of electricity to a circuit through opening or closing the switch.

Transformer - A transformer is designed to change the voltage charge within a circuit. Depending on the design of the transformer, it can either "step-up" or "step-down" the voltage output. Transformers are used with AC power.

Volt - A unit of electrical pressure which causes current to flow within the circuit.

Voltage - The force that is generated to cause the current to flow in the electrical circuit. The overall voltage is measured in volts.

Voltmeter - An instrument used to measure the force of the volts in the current. The leads are connected parallel to the points where the voltage is being measured.

Watt (Wattage) - Is a measurement for indicating the electrical power that is applied to a circuit. It can be found by multiplying the current in amperes by the volt measurement. Such as watts = amperes x volts.

Wire - A common term used to describe a conductor.

1.3 Electrical Safety

Electrical safety is an extremely important aspect of the job. Whether you are injured on or off the job, it can often lead to many struggles for you and your family. It is important that you understand the risks and how to prevent accidents while performing your duties as an electrician.

The Bureau of Labor Statistics reported that there were over 900 worker deaths on construction sites in 2016, out of which 9% were related to electrical incidents. That is over 80 deaths due to electricity. The Occupational Safety and Health Association (OSHA) has conducted studies of the major causes of deaths and injuries on construction sites, which are:

- Falls - 38%
- Struck by object - 9.6%
- **Electrocution - 8.6%**
- Caught in/between - 7.2%

It is important to remember that when on the jobsite regardless of your duties, you may be injured by many means and you should always be aware of the environment around you. In addition, OSHA has also listed the most frequent violations of safety code on construction sites:

1. Fall protection
2. Hazard communication standard
3. Scaffolding
4. Respiratory protection
5. **Control of hazardous energy (lockout/tagout)**
6. Powered industrial trucks
7. Ladders
8. Machinery and machine guarding
9. **Electrical, wiring methods, components and equipment**
10. **Electrical systems design, general requirements**

Three of the top ten violations revolve around the electrical field. Therefore, it is important that you take the correct steps by using **personal protection equipment (PPE)** to reduce the number of jobsite accidents.

Wearing the proper **personal protective equipment (PPE)** is the first step to making yourself safe. You can find the most common and necessary equipment below. Remember, you may require more or less protective clothing depending on the job you are performing or the location where the work is being done.

Head Protection - Wearing a protective hard hat will help prevent injury from falling objects and other injuries. In addition, there are several types of electrical hard hats (E,G,C) that have different ranges of protection. Make sure you know which one is required prior to entering the jobsite.

Steel-Toe Boots - Protective shoes with steel toes and rubber soles insulated against electrical shock should always be worn. These can help prevent major foot injuries that may result in permanent damage. In special cases where there are multiple rotating pieces of equipment on site, the site manager may require boots without shoestrings.

Safety Glasses/Goggles - Wearing approved eye protection helps protect your eyes from flying particles and damaging fluids. Make sure your safety glasses are approved by the ANSI. Glasses with the "+" symbol are approved for high impact.

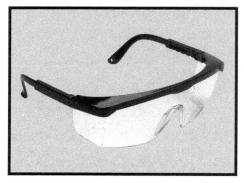

Safety Gloves - Wearing gloves can prevent injury to the hands and fingers. They can protect against burns and cuts. When working with high voltage equipment, insulated gloves can help prevent electrocution.

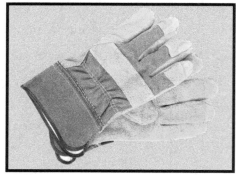

You should also **wear appropriate clothing** when on the jobsite. This includes shirts with sleeves and long pants. If it is required, you may need to wear specific fire retardant clothing. In addition, your clothes should be professional; no flaps or strings should be hanging from the clothing as they may get caught in machinery and cause injury.

The equipment you use on the jobsite should always be appropriate. Making sure the equipment you are using is up to par can mean the difference in being safe or causing an accident. **Always check equipment before entering the jobsite.**

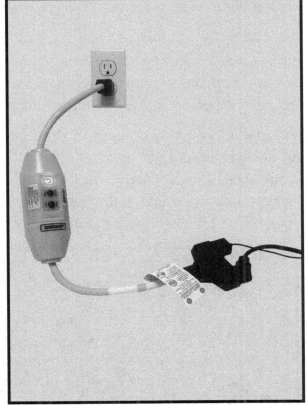

Use **insulated power and hand tools.** Ensure they are UL approved so they add protection against electric shock. Each insulated power or hand tool should have an insulation rating or resistance. Most electrical tools will have some type of insulation, either on the neck of the tool or on the handle.

Use **ground fault circuit interrupters** when you are on the jobsite. Always use a portable GFCI to protect workers on the site.

1.4 Lockout Tagout

Lockout-tagout, sometimes abbreviated as LOTO or **lock and tag**, is a process used in industry and on the jobsite that prevents the use of electrical equipment while work is taking place. The equipment used with lockout-tagout procedure can range across many industries and is an important part of safety. Before work is to be done on electrical equipment, dangerous energy from mechanical sources must be turned off and locked with the appropriate tag. This prevents any possibility of the machine being turned on accidentally.

The designated person in charge of the lockout-tagout will place a lock, sometimes multiple, on the tag which is attached to the power switch. This person will hold the keys to the lock until the job is completed and dangerous areas are clear. **If you are not the designated person to deal with the lockouts, never remove the tag yourself.**

Multiple locks are to be used in the case when there are several workers working in the designated danger area. This usually occurs on large systems where several specialists are required. Having multiple locks ensures that all workers will be clear before restoring the power to the machine. In order to use multiple locks, a clamp with several holes to secure with padlocks is placed on the switch instead of the single tag/lock.

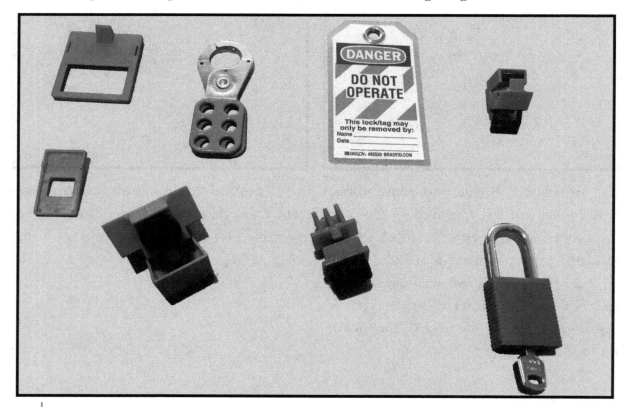

Lockout Procedure

In order to make a work area safe you must disconnect all the power sources to the equipment you will be working with. This process, also known as isolation, is necessary to meet all the safety requirements on the jobsite. The **NEC® says that the power disconnect must be within sight of the equipment being serviced.** This is to provide extra safety so that the person in charge of the lockout-tagout can see the work being done and when it is completed. The following steps should be used in lockout-tagout procedures:

1. Make a notice that the power is being shut off so all workers on the jobsite are aware.

2. Designate the energy source that is going to be disconnected in line with *NEC®* code.

3. Isolate the power source by disconnecting all energy sources used for the equipment.

4. Lock and tag all of the energy sources. If there are multiple workers then use multiple locks.

5. Double check that the locks are secure and that no energy is being transferred to the equipment.

Clamps are to be used when multiple locks are required for several workers. Clamps should always have a tag.

A **panel lockout** can prevent the panel cover from being removed. This can completely isolate the switches contained inside.

Circuit-breaker locks can be used to isolate specific circuits when other circuits must be left on, typically in residential jobs.

1.5 Electrical Tools and Symbols

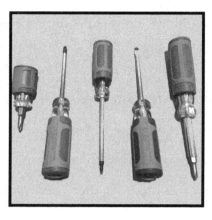

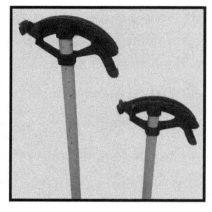

UL Approved Power Drill - Used for screw applications or hole drilling, UL approval should be listed on all power tools.

Screwdrivers - Insulated screw drivers should be used when performing electrical installations (flat head, Phillips, and Robertson).

Conduit Bender - Used to bend electrical conduit. Follow the markings on the head to attain desired angle for each bend.

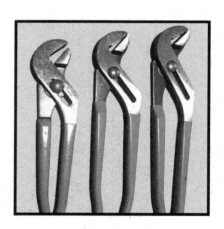

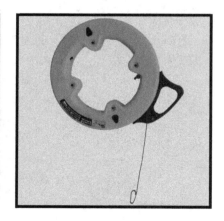

Channel Locks - Can be used to tighten locknuts or piping connections. They can be used to remove box knockouts.

Pliers - Linesman for general use, diagonal (snips) for cutting cable, and needle nose for making connections for switches and conductors.

Fish Tape - Used to guide cable by pushing or pulling wires through conduit and walls.

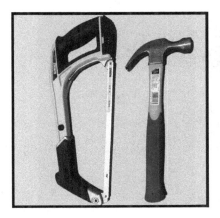

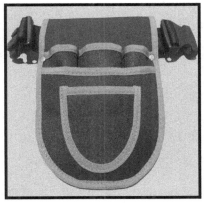

Hammer - Used to secure boxes to studs and rafters. **Hacksaw -** Used to cut large cables, PVC, or small metal pipes.

Tool Belt - Used to carry working tools. It should be heavy duty and securely strapped to your belt.

Flashlight - Used in low light conditions to maintain safe visible areas. Make sure the batteries are fresh when starting a new job.

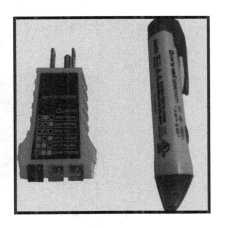

Ladder - Used to reach work that is not accessible from the ground. Make sure to use approved fiber glass or wooden ladders.

Tape Measure - Used to measure heights and lengths for switches/outlets. **Level -** Used to make sure work is level and plumb.

Live Wire Detector - Testers used by plugging into receptacles which powers light indicators to determine if there is current to the receptacle.

Ladder Safety

The Occupational Safety and Health Administration (OSHA) puts forth specific guidelines on how to properly use and secure portable ladders of all types. Falls from a ladder are a major cause of worksite injuries and can be avoided by following the guidelines below.

- Always read the labels and follow the guidelines on the ladders themselves.

- Make sure the area above the ladder is clear of obstacles and dangers.

- Make sure the ladder is in working condition with a visual inspection before use.

- Maintain a 3-point contact with the ladder at all times. This means you should have two hands and one foot, or two feet and one hand, at all times on the ladder. Never lean over one side to prevent the ladder from tipping over.

- Make sure the ladder is fully extended and not a self-supporting ladder.

- Do not place the ladder on uneven surfaces, including boxes, ramps, or any unstable base.

- Do not exceed the maximum rung usage on the ladder. Never use rungs for steps that are not designated as steps.

- Do not move or rattle a ladder which is in use.

- An extension ladder must extend 3 feet above the point of contact where the user intends to leave the ladder above ground (see diagram).

- Secure all ladders and do not exceed weight rating with equipment or tools.

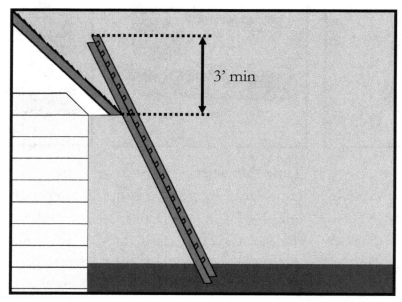

3' min

Electrical Symbols For Diagrams

Outlets		Cables and Wiring	
Single Outlet	Range Outlet	Branch Circuit	In Floor Wire
Duplex Outlet	Wall Outlet (With J-box)	Three Wire Cable	Exposed Wire
Double Duplex Outlet	Floor Outlet	Four Wire Cable	Wiring Turned Up
Ground-Fault Circuit Interrupter (Split)	Floor Duplex Outlet	In Wall Wire	Wiring Turned Down
Weatherproof Outlet	Fan Outlet	Overcurrent Device	
Arc Fault Circuit	Ceiling Outlet (With J-box)	Single Pole Breaker	Double Pole Breaker (Tie Handle)
GFCI Protected Outlet	Special Purpose Outlet	Fuse	
Split Wire Duplex Outlet	Recess Fixture Outlet	Disconnects	Lighting Outlets
Triplex Outlet	Ceiling Fan Outlet	Fusing Not Possible	Ceiling Outlet
Dryer Outlet	Vent Fan Outlet		Fluorescent Fixture
Switches			Long Row Fluorescent Fixture
S — Single Pole Switch	S_F — Fan Switch	Can be Fused	
S_2 — Double Pole Switch	S_P — Switch With Pilot Light		Other Symbols
S_3 — Three Way Switch	M — Motor Switch		Electrical Panel
S_4 — Four Way Switch	T — Time Switch		Telephone Jack

1.6 **Electrical Voltage Testing**

Learning to properly read and measure the electrical volt/ohm output is an important part of your job as an electrician. Electrical meters come in various sizes and can provide you with various readings such as amps, volts, ohms, continuity, hertz, and capacitance. Volt/Ohm meters are rated by categories ranging from I - IV. IV is the highest rated and will provide the highest voltage and energy transients (voltage spikes) rating. We have provided a multimeter numbered guide for your convenience.

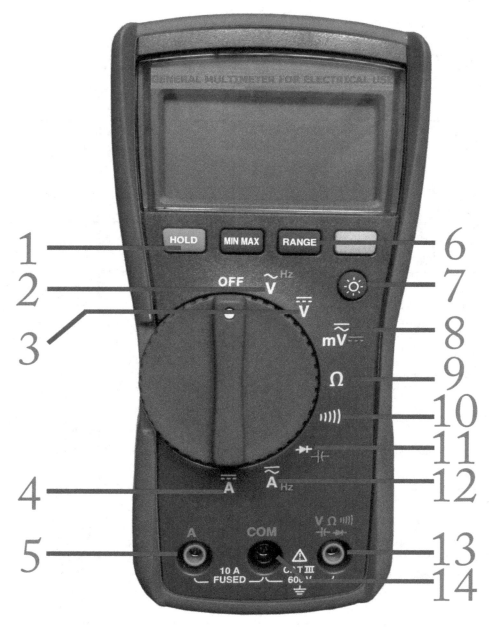

1. **Hold.** This button will pause the reading on the screen so that you can reference it later. This is useful if you are testing an area where viewing the screen of the meter is difficult. Simply press the "hold" button to use this feature.

2. **AC Voltage.** This voltage setting will provide you with a voltage reading for the tested circuit. This is one of the most common measurements used. **SHIFT** can also be used to measure the frequency of your voltage reading. This is important to know because many systems are designed with fluctuating frequencies.

3. **DC Voltage.** This setting is primarily used to test small electronic devices, batteries, or small LED lights. DC voltage readings are typically very low.

4. **Direct Current.** Normally performed with a clamp attachment. This test will show you the direct current load applied to small plugged in devices.

5. **Current Jack.** Used with a red test lead or clamp for current only, not to be used with black test lead.

6. **Range.** Can be used to change the range of the meter. Some meters do this automatically but you can use the range button to cycle between volts or similar ranges.

7. **Brightness.** Changes the brightness of the screen to make it easily visible.

8. **AC Millivolts.** This setting is used when testing small circuits. If you need more accurate readings than the normal AC setting, the millivolt setting will provide a more precise measurement. Press **SHIFT** to change to DC millivolts.

9. **Ohms.** This setting is used to measure resistance, also known as ohms. It is normally used to test the condition of wires. It can display several readings. If the reading says "OL" then there is a faulty fuse. ONLY test fuses when they are taken out of the circuit.

10. **Continuity.** This setting provides an audible tone when it detects continuity between your black and red test leads. This setting can be used to find open and short circuits quickly.

11. **Diode Test.** This setting is used to check the status of individual component diodes. It can accurately detect if the diodes are good or bad.

12. **Alternating Current.** Similar to direct current (number 4) except with alternating.

13. **Red Jack.** Red wire testing for all readings besides current (number 5).

14. **Common Jack.** Used with the black wire lead for all tests.

1.7 OHM's Law

Ohm's Law is the method used to determine voltage, ohms, watts, or amps based on equations within Ohm's Law. It states that the current through a conductor between any two points is the same as the voltage across those two points as well. When we calculate values using Ohm's Law, we use the following symbols to represent our electrical readings:

I = Current Intensity = **Amperes**

E = Electromotive Force = **Volts**

P = Power = **Watts**

R = Resistance = **Ohms**

There are three basic equations that can be used with Ohm's Law to determine a missing value. Make sure that when you are using these equations that your units for each value are the same.

Ohm's Law formulas:

$$R = \frac{E}{I} \qquad I = \frac{E}{R} \qquad E = I \times R$$

This is a **universal chart** used to find the equations you need with Ohm's Law. In order to use this chart, **look for what value you need (amps, watts, volts, or ohms) and then look at that corner section for the correct equations to use.**

This is a very useful tool to use in the field as it can be quickly referenced when calculations need to be done on the fly.

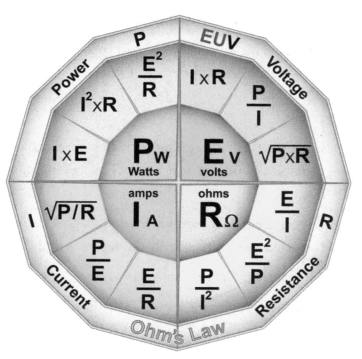

Example:

A hair dryer that is connected to a receptacle is rated at 960 watts and is drawing 8 amperes. What is the voltage being supplied to the hair dryer?

$$Volts = \frac{Watts}{Amperes} \qquad E = \frac{P}{I} \qquad E = \frac{960}{8} = 120\ V$$

What is the resistance in this case?

$$Ohms = \frac{Volts}{Amperes} \qquad R = \frac{E}{I} \qquad R = \frac{120}{8} = 15\ \Omega$$

With these numbers, we can then use the other Ohm's Law equations to calculate any value we need.

POWER

$$Watts = \frac{Volts^2}{Ohms} \qquad P = \frac{E^2}{R} = \frac{120^2}{15} = \frac{14,400}{15} = 960\ W$$

$$Watts = Amps^2 \times Ohms \qquad P = I^2 \times R = 8^2 \times 15 = 960\ W$$

$$Watts = Volts \times Amps \qquad P = E \times I = 120 \times 8 = 960\ W$$

CURRENT

$$Amps = \frac{Watts}{Volts} \qquad I = \frac{P}{E} = \frac{960}{120} = 8\ A$$

$$Amps = \sqrt{\frac{Watts}{Ohms}} \qquad I = \sqrt{\frac{P}{R}} = \sqrt{\frac{960}{15}} = 8\ A$$

$$Amps = \frac{Volts}{Ohms} \qquad I = \frac{E}{R} = \frac{120}{15} = 8\ A$$

RESISTANCE

$$Ohms = \frac{Volts}{Amps} \qquad R = \frac{E}{I} = \frac{120}{8} = 15\ \Omega$$

$$Ohms = \frac{Watts}{Amps^2} \qquad R = \frac{P}{I^2} = \frac{960}{8^2} = 15\ \Omega$$

$$Ohms = \frac{Volts^2}{Watts} \qquad R = \frac{E^2}{P} = \frac{120^2}{960} = 15\ \Omega$$

VOLTAGE

$$Volts = \frac{Watts}{Amps} \qquad E = \frac{P}{I} = \frac{960}{8} = 120\ V$$

$$Volts = Amps \times Ohms \qquad E = I \times R = 8 \times 15 = 120\ V$$

$$Volts = \sqrt{Watts \times Ohms} \qquad E = \sqrt{P \times R} = \sqrt{960 \times 15} = 120\ V$$

1.8 Structure Framing

It is important that you understand the basics of structure framing when installing electrical systems. You will often hear terms referring to the frame structure that you should easily recognize. Framing plans may be dramatically different but all contain some of the same basic framing parts in typical wood or metal framing designs.

The parts below refer to the diagram on the following page:

1. **Rafters** - Form the roof support structure while being supported by the ceiling joists and top plate.

2. **Ceiling Joists** - The horizontal framing structure that is attached to the top plate and sits on top of the wall framing studs.

3. **Top Plate** - The top portion of the horizontal section of the framework. It is usually made of 2" x 4" or 2" x 6" wood beams.

4. **Wall Studs** - Form the vertical support structure for the framework. It is usually made of 2" x 4" or 2" x 6" wood beams.

5. **Bottom Plate** - The bottom portion of the horizontal section of the framework. It sits on the subfloor. It is usually made of 2" x 4" or 2" x 6" wood beams.

6. **Subfloor** - The first solid layer of material used to cover the floor joists. It is usually made of 4' x 8' plywood sheets.

7. **Floor Joists** - The horizontal framing structure that is attached to the sill plate and supports the floor and wall structure stability. It is usually made of 2" x 8", 2" x 10", or 2" x 12" wood beams.

8. **Sill Plate** - Wood plate structure used to secure the floor joists. Sill plates sit on top of the foundation and are usually fastened with J-bolts.

9. **Foundation** - The base of the structure that the rest of the framework will be built upon.

10. **Footing** - The below ground concrete base that provides support to the foundation.

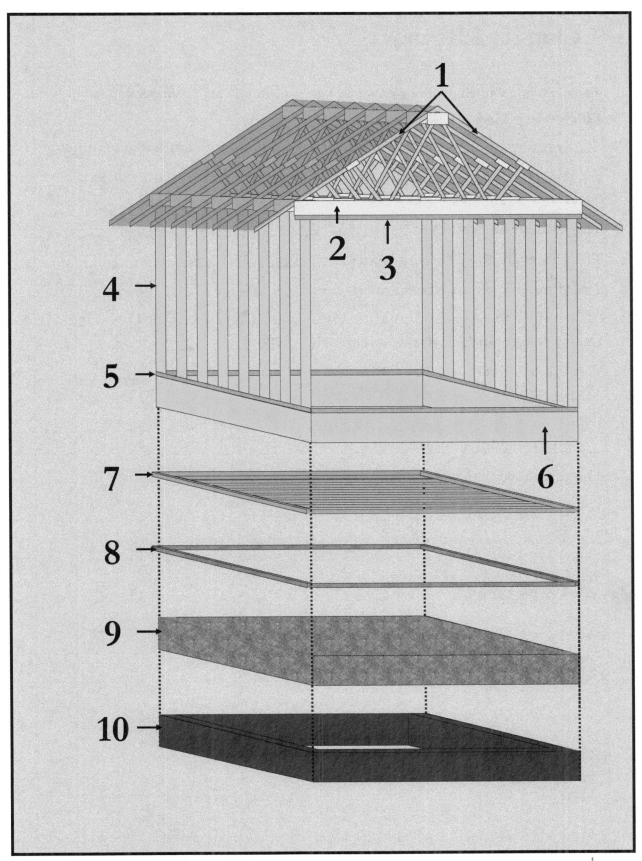

1.9 Chapter 1 Review

1. Two ways electrical energy can be produced are __Produced Coal__ and
   __Natural gas__ __renewable resources__

2. __electricity__ is the flow of electrons from atom to atom in a conductor.

3. __pigtail (jumper)__ is a short wire or conductor used to connect two or more wires to a single screw terminal.

4. The letters PPE stand for __Personal protective equipment__.

5. The abbreviation LOTO translates to __Lockout - tagout__.

6. The three types of screwdrivers are __Flat head__, __phillips__, and __roberdson__.

7. Draw the symbol for a duplex receptacle __⊖__.

8. Volt/Ohm meters are rated by categories ranging from __I__ to __IV__.

9. Ohm's Law is a method used to determine __Voltage__, __ohms__, __Watts__, or __amps__.

10. The three basic Ohm's law formulas are:

    __$R = \frac{E}{I}$__, __$I = \frac{E}{R}$__, __$E = I \times R$__

CHAPTER 1 NOTES:

Sketch #1

Electrical Boxes

2

Electrical boxes act as the housing for conductor or wire connections in order to protect both workers and the connection from damage. Electrical boxes can also secure lighting or fan fixtures through attachment to supports or ceiling beams.

Boxes come in a variety of shapes, sizes, colors, and materials. The type of box that needs to be installed will depend on the wiring plan and its intended installation location.

Boxes can be waterproof or provide grounding for the conductors. Some electrical boxes come with bar hangers so they can securely mount heavy fixtures.

Always follow *NEC®* regulations when installing boxes. There are specific requirements for the installation area, material, and spacing within the box depending on the size wire being housed within them.

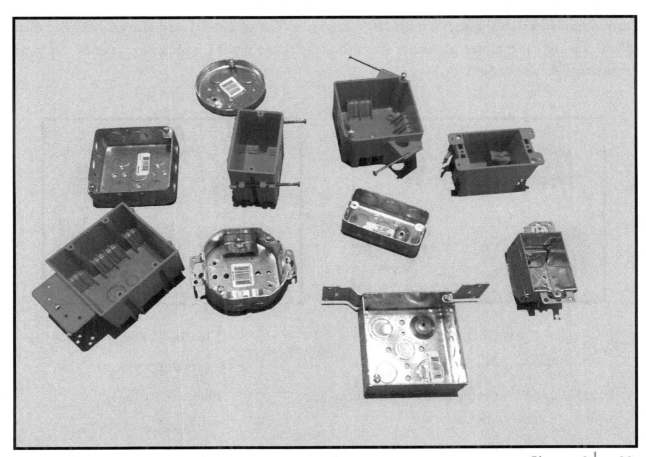

2.1 Metal and Nonmetal Boxes

Electrical boxes are an extremely important part of your residential or commercial electrical system. Choosing the right box can be overwhelming. Boxes come in several styles, materials, sizes, and shapes. The two main types of boxes you will usually choose from are plastic or metal boxes. These boxes can be used for ceiling and wall lighting, fan fixtures, and junction boxes.

When choosing a box you should consider the following:

- What the code requirements are

- Which boxes you have more experience with

- Which items you have on hand that meet your budget

- The intended use of the box

There are a variety of styles of boxes, such as new work and old work, round, square, and octagonal boxes. The depth of these boxes can range from 1/2" to more than 3". If you are **using metal conduit pipe to run the wiring then a metal box is required** due to grounding through the metal itself. If you plan to use nonmetal wiring type NM-B cable, then you may use either plastic or metal boxes. These types of situations usually apply to residential establishments.

Metal Boxes

Drawn Steel (1 piece)

Interlocked / Gangable (used for switch and outlet boxes)

Plastic Boxes

PVC (most common)

Phenolic Resin

Fiberglass

Electrical boxes for either the wall or ceiling can be used in a variety of ways, including:

- to mount and hold connections for light fixtures to a wall or ceiling

- to mount and hold connections to a ceiling fan

- to join or splice wire in circuits

- can be mounted, recessed in the wall, or on the surface

The next section will provide examples of commonly used boxes and their typical applications. Remember that no box is universal and careful consideration should be given when selecting a box for a project.

Round Pan Electrical Box (Pancake)	**4" Square Box**

Round Pan Electrical Box (Pancake)

Size

1/2" and 3/4" Deep; 3-1/2" Round

When to use:

- The metal round pan box may be used for mounting lights to the wall or ceiling that weigh less than 50 pounds.

- The plastic round pan box may be used to mount outlets.

- Certain variations of round pan electrical boxes may be used for fans but are not typically used.

4" Square Box

Size

1-1/4" to 2-1/8" Deep

When to use:

- Square boxes are most often used to run several conductors in multiple directions.

- They allow the most volume for placement of several conductors.

- Square boxes can also be used when installing wall or ceiling lighting as well as switches and receptacles.

Octagon Electrical Box	**Ceiling Fan Rated Electrical Box**

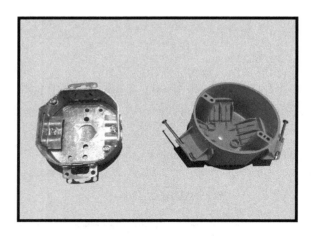

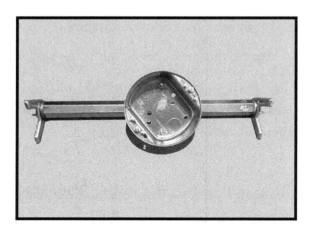

Size

1/2" to 2-1/8" Deep

When to use:

- Octagon type boxes can be used as junction boxes.

- The plastic octagon box typically has "ears" for connection to the wall or ceiling when being used for "old work" jobs.

- The metal round or octagon box can be used for an outlet with the use of NM cable or conduit.

Size

1/2" to 2-1/8" Deep

When to use:

- The shape of these boxes may be a round or octagon box approved by UL.

- They require specific fastening and mounting due to the rotating fan.

- The boxes must be rated by UL for use with ceiling fans, typically these boxes will say "for use with ceiling fan".

- These boxes may be approved for fan installations up to 75 pounds or more. Check the specific box for weight restrictions.

Metal Boxes

Metal boxes have different installation procedures than plastic boxes. Metal boxes require additional steps to installation and grounding of devices housed in the box. The following are important parts of the metal box:

1. **Device Mounts** - Used to secure switches and receptacles to the box.

2. **Mounting Ears** - Used to easily install the box and adjust it to the right position.

3. **Knockouts** - Small quarter size plugs located on the sides, top, and bottom of the box that can be removed to connect electrical cables and feed conductors into the box.

4. **Grounding Screw** - The grounding screw is only found in metal boxes and is used to ground the equipment housed in the box with a grounding conductor attachment. The **NEC®** states that all devices in metal boxes must be grounded this way.

5. **Pry Outs** - Provide access inside the box for installing other cables.

6. **Cable Clamps** - Found only in metal boxes to secure cables to the box.

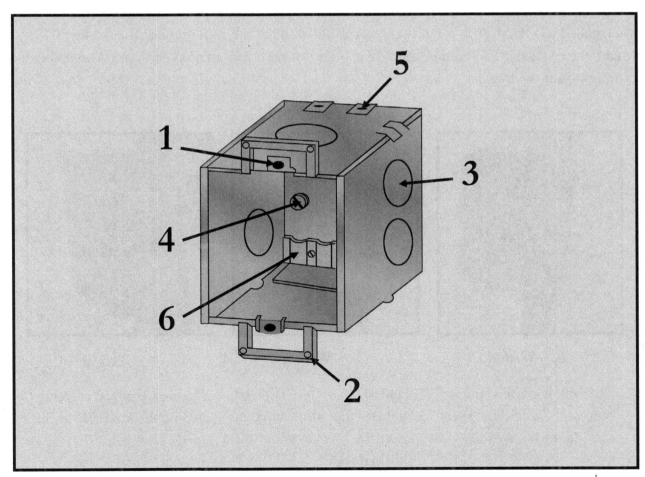

2.2 **Installing Boxes**

Electrical boxes can be installed in many locations depending on where they are needed. Some boxes are nailed on while others may need to be mounted a different way. When it is time to install the box, the height will be indicated by a mark on the stud where it is to be mounted. Different businesses may have different ways of marking and positioning the box, but we suggest using the following:

For **square boxes**, mount the **bottom of the box on the mark.**

For **round boxes**, mount the **center of the box on the mark.**

No matter what type of box you are using, it is important to use the right depth when installing the box. The depth marks on the side of the box should be flush against the stud when the box is being set with a correct depth for 1/2 inch sheet rock. If the sheetrock is set to be thicker than normal, such as with a firewall, then the box placement can be pulled out farther to accommodate the change.

As a rule of thumb, it is always better to install the box too deep than too far out. If the box is installed too deep, then longer screws can be used for the mounting of the devices. If you make the mistake of mounting the box too far out then you will have to cut the install nails and remount the box.

X - Wrong

The box is placed too far out, and would be seen outside the sheetrock.

X - Wrong

The box is not square against the stud and it could be seen outside the finished wall.

✓ - Right

The box is place correctly against the stud and at the correct depth for 1/2 inch sheetrock.

Nonmetal Box Installation

Installation of nonmetal boxes are usually done by securing them against studs using nails or screws. In some cases the boxes may be secured to the studs using brackets.

Determine which knockouts in the box you want to use for the cables. Open them using a screwdriver. **Unlike metal boxes, there are no pry-outs and cables will always be run through knockouts for nonmetal boxes.**

After determining the position of the box on the stud, use the appropriate mounting device (screw/nails or mounting bracket) to secure the box.

Check to make sure that the depth of the box is correct and that it will line up with the inside wall.

Metal Box Installation

Metal boxes are installed similar to nonmetal boxes. **The only difference here is the decision on whether the cables will enter through the knockouts or the pry outs.** If you are using the knockouts they should be removed prior to securing the box.

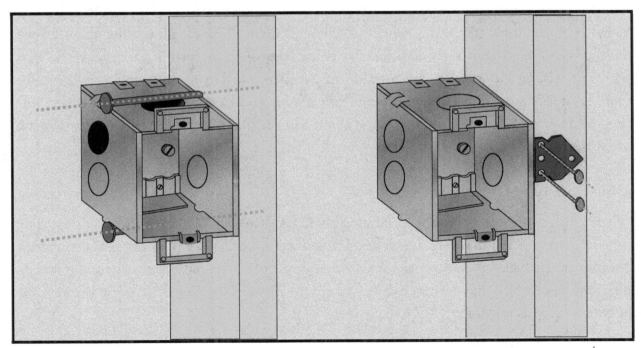

You will sometimes have to mount a box in narrow spaces. These small spaces, which are typically between two studs that lie less than 6 inches apart, make it impossible to swing a hammer to mount the box. There are a few options if you must install a box in a narrow space:

1. Use a front mount box that does not require side nails.

2. Mount the standard box but use wood screws in the place of nails.

When installing the outlet box against the ceiling joist you should nail it to the joist through the predrilled holes. **Always consider the thickness of the drywall and place the box where it will be flush with the outside of the drywall.** If you are installing a box when there is already sheetrock present, then it may require you to go into the attic to secure the box to the joist. Make sure there is room to drill a hole below the box in the drywall to run the wires.

Adjustable Bar Hangers

When you are installing overhead lighting or fan fixtures, it is usually done by installing adjustable metal bar hangers between ceiling joist in order to provide added support. Adjustable bar hangers can be used with either metal or nonmetal boxes. There are several different types of bar hangers including straight or offset, adjustable or fixed, and can come in several lengths including 18, 24, or 30 inch.

The following are the steps to installing an adjustable bar hanger:

1. Assemble the adjustable bar hanger with the outlet box if it was not pre-installed. Follow the instructions given by the manufacturer of the bar hanger.

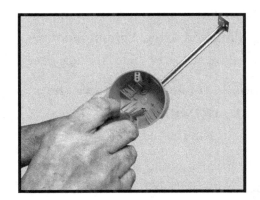

2. Adjust the box to ensure it is positioned correctly and is at the correct depth. Make sure the box is flush with the ceiling. Secure the hanger in place by lightly hammering it between the ceiling joists.

3. Once the hanger is in the correct place, hammer in the nails or use screws to attach the bar hanger to the ceiling joists.

4. Once the bar hanger is fixed, adjust the box to the correct position for the fixture install. Tighten the set screw on the bar hanger to prevent the box from moving.

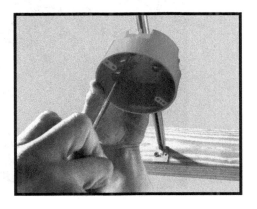

<u>Heavy Load Rated Boxes</u>

The *NEC®* has specific requirements regarding the installation of heavy fixtures such as ceiling fans and large lighting fixtures. To ensure you do not violate any codes, heavy fixtures should not be held in place by the box alone.

The box you use should have the stamp of approval from Underwriters Laboratory (UL) for use with heavy fixtures. If UL has given the approval for a box to be the sole support for the fixture, then it may be used.

Many popular fan and lighting fixture manufacturers provide support systems that meet the requirements put forth by the *NEC®*. They may be either metal or nonmetal boxes. They are installed using special brackets between the ceiling joists. You should follow the instructions given by the manufacturers for installations.

2.3 Box Fill

The subject of box fill is often overlooked or passed over quickly. However, if you ask seasoned electricians where they encounter the most violations, they almost always say box fill requirements. It is common that boxes are often used with more fittings and devices than intended. This violation happens too often and in this section you will learn how to prevent a box fill violation.

The danger of over-filling a box may not be obvious, but when boxes are overpacked it can create a serious fire hazard. When boxes are filled with load carrying conductors, heat is produced. To reduce the heat in the box there must be enough space to allow the heat to vent properly. In addition to adding heat by having additional equipment in the box, you also block the heat from escaping. Although the danger is not immediate, over time the heat can build and cause electrical fires. The *NEC®* contains rules and requirements that deal with box fill. For more information on box fill, refer to *NEC®* **Article 314.**

General Rules for Box Fill

There are specific sizing requirements for boxes that contain devices or utilization equipment. For boxes that do not contain these devices, they are permitted to have an internal depth of less than 12.7 mm (1/2 in.) For all other boxes, the requirements can be seen in the table below. For more information regarding box sizing and heat restrictions please refer to *NEC®* **314.25 and 410.25.0**

- When large depth boxes are required (1 7/8 in. rearward or greater), then there must be at least 1/4 in. depth before reaching equipment contained in the box.

- Specific box types or sizes may be required if you are installing special devices or utilization equipment.

Conductor Size Being Installed	Minimum Depth of Box
More than 4 AWG	Refer to specified reference depth
4, 6, or 8 AWG	2 1/16 in.
10 or 12 AWG	1 3/16 in.
14 AWG or Smaller	15/16 in.

The total volume of the box is what determines the number of conductors or devices that can be contained within it. There is a variety of equipment that can occupy the box, including conductors, fittings, devices, and receptacles. Each piece of equipment (clamps, receptacles, conductors, etc.) has its own specific requirement for free volume space in the box. Space requirements for conductors can be found in either cubic inches or cubic centimeters in *NEC®* Table 314.16(B). When referring to this table, the box must provide at least this minimum volume requirement.

Size of Conductor (AWG)	Free Space Within Box for Each Conductor	
	cm³	in.³
18	24.6	1.50
16	28.7	1.75
14	32.8	2.00
12	36.9	2.25
10	41.0	2.50
8	49.2	3.00
6	81.9	5.00

NEC® table 314.16(B) - Reprinted with permission from the National Electrical Code® Handbook 2017, copyright © 2017 National Fire Protection Association. All rights reserved

If the box to be installed provides one or more barriers, then the volume is distributed to each of the spaces within. Most barriers will have a marking with its volume. If the barrier does not show a specific volume, you should assume that metal boxes will occupy a volume of 1/2 in.³ and 1 in.³ if it is nonmetal. The volume considerations were not a concrete code before, however they are officially documented in the 2017 *NEC®* Handbook (*NEC®* 314.16(A)). The spaces contained within the box installed with a barrier must also be considered. Each volume within boxes installed with barriers must be calculated to ensure proper volume for the contained equipment. This can be found in *NEC®* 314.16(B).

Box Volume Requirements - NEC® table 314.16(A)											
Box Trade Size			**Minimum Volume**		**Maximum Number of Conductors** (arranged by AWG size)						
mm	in.	Box Type	cm³	in.³	18	16	14	12	10	8	6
100 x 32	4 x 1 1/4	Round/ octagonal	205	12.5	8	7	6	5	5	5	2
100 x 38	4 x 1 1/2	Round/ octagonal	254	15.5	10	8	7	6	6	5	3
100 x 54	4 x 2 1/8	Round/ octagonal	353	21.5	14	12	10	9	8	7	4
100 x 32	4 x 1 1/4	Square	295	18.0	12	10	9	8	7	6	3
100 x 38	4 x 1 1/2	Square	344	21.0	14	12	10	9	8	7	4
100 x 54	4 x 2 1/8	Square	497	30.3	20	17	15	13	12	10	6
120 x 32	4 11/16 x 1 1/4	Square	418	25.5	17	14	12	11	10	8	5
120 x 38	4 11/16 x 1 1/2	Square	484	29.5	19	16	14	13	11	9	5
120 x 54	4 11/16 x 2 1/8	Square	689	42.0	28	24	21	18	16	14	8
75 x 50 x 38	3 x 2 x 1 1/2	Device	123	7.5	5	4	3	3	3	2	1
75 x 50 x 50	3 x 2 x 2	Device	164	10.0	6	5	5	4	4	3	2
75 x 50 x 57	3 x 2 x 2 1/4	Device	172	10.5	7	6	5	4	4	3	2
75 x 50 x 65	3 x 2 x 2 1/2	Device	205	12.5	8	7	6	5	5	4	2
75 x 50 x 70	3 x 2 x 2 3/4	Device	230	14.0	9	8	7	6	5	4	2
75 x 50 x 90	3 x 2 x 3 1/2	Device	295	18.0	12	10	9	8	7	6	3
100 x 54 x 38	4 x 2 1/8 x 1 1/2	Device	169	10.3	6	5	5	4	4	3	2
100 x 54 x 48	4 x 2 1/4 x 1 7/8	Device	213	13.0	8	7	6	5	5	4	2
100 x 54 x 54	4 x 2 1/8 x 2 1/8	Device	238	14.5	9	8	7	6	5	4	2
95 x 50 x 65	3 1/4 x 2 x 2 1/2	Masonry box	230	14.0	9	8	7	6	5	4	2
95 x 50 x 90	3 1/4 x 2 x 3 1/2	Masonry box	344	21.0	14	12	10	9	8	7	4
min. 44.5 depth	FS - single cover (1 3/4)		221	13.5	9	7	6	6	5	4	2
min. 60.3 depth	FD - single cover (2 3/8)		295	18.0	12	10	9	8	7	6	3
min. 44.5 depth	FS - multiple cover (1 3/4)		295	18.0	12	10	9	8	7	6	3
min. 60.3 depth	FD - multiple cover (2 3/8)		395	24.0	16	13	12	10	9	8	4

Situations may arise where the box you are using does not meet the sizing requirements for the conductors and devices housed in the box. **In order to avoid any volume violations, you can install an extension ring**. The ring must be the same shape as the installed box.

Box Sizing Examples

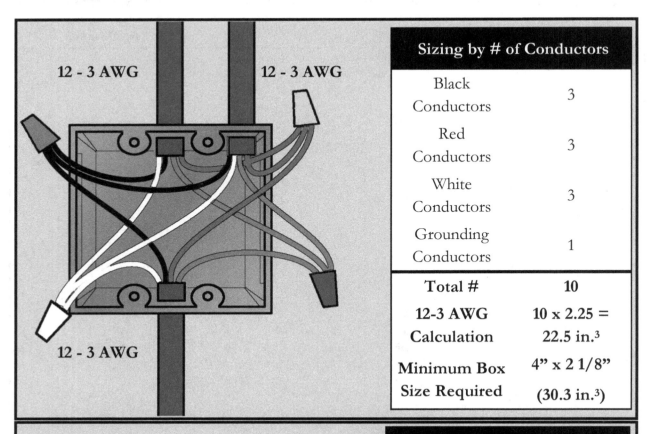

Sizing by # of Conductors	
Black Conductors	3
Red Conductors	3
White Conductors	3
Grounding Conductors	1
Total #	**10**
12-3 AWG Calculation	**10 x 2.25 = 22.5 in.^{3}**
Minimum Box Size Required	**4" x 2 1/8" (30.3 in.^{3})**

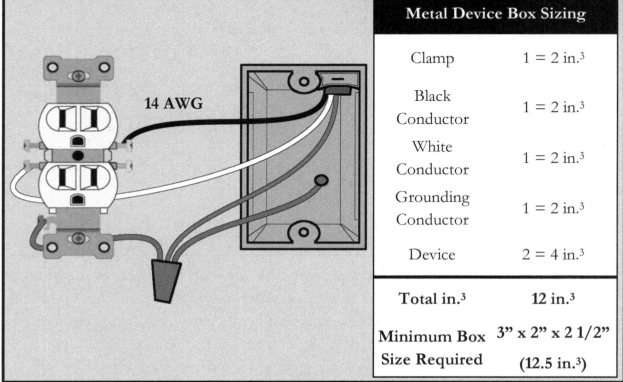

Metal Device Box Sizing	
Clamp	1 = 2 in.^{3}
Black Conductor	1 = 2 in.^{3}
White Conductor	1 = 2 in.^{3}
Grounding Conductor	1 = 2 in.^{3}
Device	2 = 4 in.^{3}
Total in.^{3}	**12 in.^{3}**
Minimum Box Size Required	**3" x 2" x 2 1/2" (12.5 in.^{3})**

2.4 J-Box Sizing

There are specific guidelines when sizing junction boxes, pull boxes, handhole enclosures, and conduit bodies. These guidelines are put in place to prevent damaging of the conductor insulation. The specific requirements can be found in **NEC® code 314.28** and applies to all conductors of #4 AWG and larger.

Calculation of Pull Box Enclosures

Straight Pull - Multiply the largest raceway size by 8. This will give you the minimum length for the box.

Example:

raceway of 2 inches ⟶ 2" x 8 = 16"

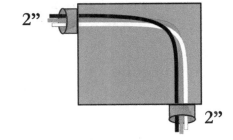

Angle Pull - Multiply the largest raceway size by 6, but then add the sizes of all other raceways that are on the same wall and row to get the minimum box size.

Example:

raceway of 2 inches ⟶ 2" x 6 = 12"

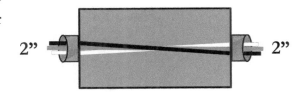

U-Pull - Calculate the same way as an angle pull.

Example:

raceway of 2 inches ⟶ 2" x 6 = 12"

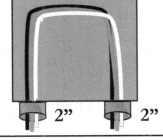

Multiple - Calculate each one separately by row.

Example:

raceway of 3-2" and another that is 3-1"

Row 1 ⟶ (2" x 6) + 2" + 2" = 16"

Row 2 ⟶ (1" x 6) + 1" + 1" = 8"

Choose 16" because it is larger.

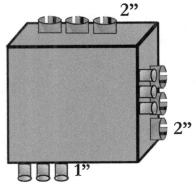

Distance Between Raceways - The distance between raceways in the box that hold the same conductor cannot be less than six times the largest raceway. It is measured from the closest edge to edge.

Example:

14" distance from edge to edge

raceway of 2 inches $\longrightarrow$ 2" x 6 = 12"

14" > 12" = safe to use

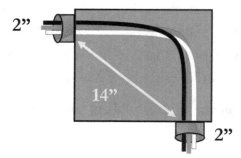

Steps to Pull Box Calculation:

When determining the correct box size, you will have to calculate both the horizontal and vertical dimensions. You will always have to measure the distance between raceways as well to make sure the distance from edge to edge is acceptable.

For our **example**, we will assume our pull box has multiple raceways, 1-3" raceway on the left, 1-3" raceway on the right, 1-2" raceway on the left, 1-2" raceway on the bottom.

Horizontal Dimension

Straight Pull:

- Left to Right: 8 x 3" = 24"
- Right to Left: 8 x 3" = 24"

Angle Pull:

- Left to Right: (6 x 2") + 3" = 15"
- Right to Left: No Measurement

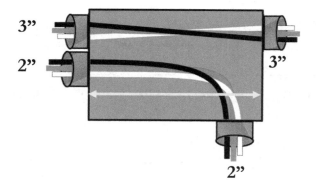

Vertical Dimension

Straight Pull:

- No Measurement

Angle Pull:

- Top to Bottom: No Measurement
- Bottom to Top: 6 x 2" = 12"

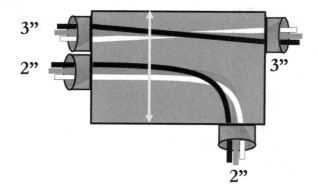

Distance Between Raceways

- 2" x 6 = 12"

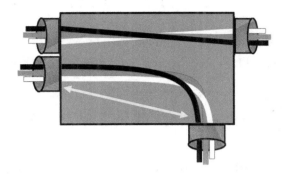

Therefore, your final box must be 24" wide and 12" tall.

You may be asking, what is the point of using a straight pull when wires will not be switching directions. Would it not be easier to not break the conductor piping with the addition of a box? There are multiple reasons why you add a straight pull:

- The straight pull box provides a lubrication point.
- Less force is needed to run the wires with a pull through walls or piping.
- It helps support long conductor runs.
- It can reset the limit on the 360° bend rule for raceway piping.

Remember that the box must meet the minimum requirements based on these calculations. Make sure to double check your calculations and that the box you are using meets these requirements. A better option would be to have a fellow electrician or supervisor check your calculations.

There are a few special rules when determining the required volume for a box. Some fixtures come equipped with a wire housing cover which can hold conductors if they are no larger than #14 AWG wire. Since these conductors are not directly housed in the box, they do not need to be included in the volume calculations.

2.5 Chapter 2 Review

1. The two main materials used in electrical boxes are _____ and _____.

2. When installing fans or overhead lighting, the boxes are usually supported by _____.

3. When boxes are overfilled with conductors it can produce too much _____.

4. Space requirements for conductors in boxes can be found in *NEC®* table _____.

5. To avoid violations of volume requirements, you can install a _____ to increase the volume of the box.

6. Always consider the _____ of the wall covering before installing boxes.

7. _____ boxes are used for mounting lights where the depth of the outlet is no more than 1/2" to 3/4".

8. For angle pulls, you would multiply the largest raceway in the box by _____ .

9. For straight box pulls, you would multiply the largest raceway by _____ .

10. When determining box size you must calculate both _____ and _____ dimensions.

CHAPTER 2 NOTES:

Electrical Cables and Conductors 3

The term "cables" originated as a name for ship line where several ropes are woven together to create a stronger and thicker line to anchor large boats. As the needs of the electrical industry continued to expand, stronger wires were needed for long distance conductor lines. Therefore, these wires were woven together and thus electrical cables were born. These bare copper wires are woven together with a type of sheath that covers the exposed cable.

The term wire and cable are sometimes referred to as the same thing, but a cable is a combination of multiple wires sheathed together. These cables usually connect two or more devices to transfer electrical power between them, and are made in different sizes depending on the application. Wires are used to carry a current from the power source to a receptacle or piece of equipment such as a light or fan. When someone mentions a conductor or wire, they usually mean the same thing.

Whether you are using cables or wires, proper installation is extremely important to deliver steady and safe electrical power. Proper installation will also ensure that your wiring will pass electrical inspection. When installing these wires and cables, you should follow the National Electrical Code (*NEC®*) and local regulatory codes to determine proper usage of each type of wire or cable in your wiring plan. Before you begin the installation, make sure you have received approval from the building inspector, and have it inspected after you have finished the job.

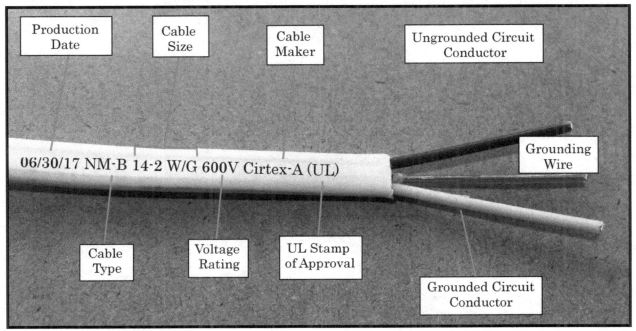

Production Date | Cable Size | Cable Maker | Ungrounded Circuit Conductor

06/30/17 NM-B 14-2 W/G 600V Cirtex-A (UL)

Grounding Wire

Cable Type | Voltage Rating | UL Stamp of Approval | Grounded Circuit Conductor

3.1 Cable Types

Electrical cables are produced in several different types and sizes depending on their application. They are used to provide services such as internet or landlines up to large scale industrial use. There are two main categories for cables based on their sheathing type. These two types are metal and nonmetal sheathing.

Metal Sheathed Cables

1. MC Cable

These cables, often referred to as armored or MC cables, are sheathed in some type of metal. They are used to power electrical lines for appliances such as commercial refrigerators or ovens. These cables usually contain three or four plain solid copper wires for the ground, current, and neutral placement. These wires are usually insulated with a polyethylene blend used as a coating. MC cables are often used when the wires need to hold up to heavy duty use or strain, and are often implemented in outdoor situations. MC can be installed outdoors, but wires cannot terminate outside.

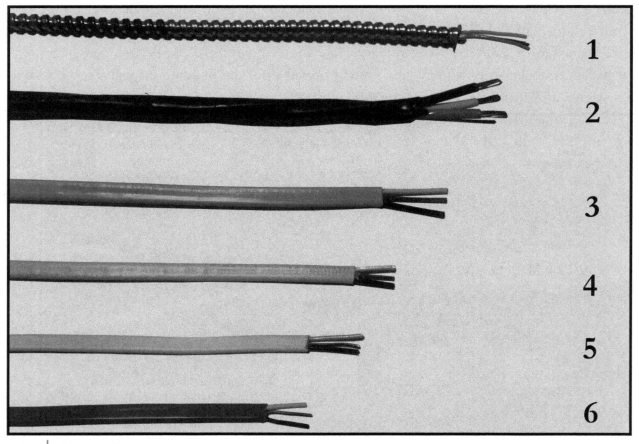

Nonmetal Sheathed Cables

Nonmetal cables (NM) have a flexible plastic sheath that contains 2-4 wires coated with a heat resistant plastic and one bare grounding wire. These cables come in various colors and sizes depending on their application. The most common cables are NM-B and NM-C which are used indoors. They come in white, yellow, orange, black, or gray, and are all solid color insulation. Black is used twice but with different gauge sizes, so be careful when using black cable that you are selecting the appropriate gauge.

2. Black Color Cable

Black color coded cable can come in several different gauge sizes and is typically found in either 8 or 6-gauge rated wire. The 8-gauge wire can be used for 40-amp circuits and the 6-gauge wire can be used for power up to 55-amps. The 8-gauge wire is used when it meets requirements, however the 6-gauge wire is preferred for subpanel feeds or large industrial appliances or equipment.

3. Orange Color Cable

The orange color coded cable contains 10-gauge wire that is used for 30-amp circuits for a range of large applications. They are typically used for dryers, water heaters, and any other appliance loads that require this amperage.

4. Yellow Color Cable

The yellow color coded cable contains 12-gauge wire that is used for 20-amp circuits for the home or small business. They are typically used for outlets and household appliance power.

5. White Color Cable

The white color coded cable contains 14-gauge wire that is used for 15-amp circuits in the home or small business. They are used for small light and fan fixtures.

6. Grey Color Cable

Gray colored cables have the widest range of uses. Gray colored cables are put in categories based on their use and not on their size or gauge.

Below-ground Feed Cable

These gray cables come in several sizes and are designated as underground feeder (UF) cables. They are similar to the traditional NM cables but they are coated in a thermoplastic and sheathed in a flexible material to protect the individual wires. This makes them highly resistant to outdoor conditions.

Coaxial Cable

This gray cable contains several layers of conductors and shielding in a tubular pattern. They are often used to supply a steady signal for digital audio or cable television. The conductors within the coax cable are spaced and insulated equally.

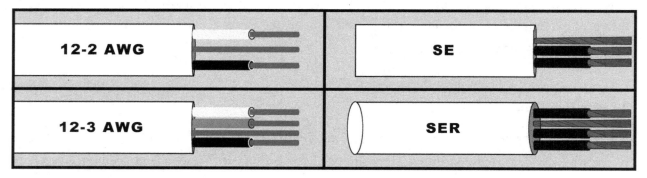

NM Types

12-2 AWG - 3 wire cable with 2 conductors and one grounding wire.

12-3 AWG - 4 wire cable with 3 conductors and one grounding wire.

Service Entrance Cables

SE - Service Entrance cable for normal service entrance jobs, can be used without sheath.

SER - Round Service Entrance cable with 3 insulated wires and a grounding wire.

Power Cords

Power cords come with a variety of features depending on the requirements of the job. Each portable power cord comes with a set of letters designating what it is used for:

S - Stands for "Service" cable and is rated up to 600 volts.

SJ - Stand for " Junior Service" where the cord is rated at 300 volts.

O - Is given when an oil resistant outer jacket is on the cord.

W- Stands for water and weather resistant and usually comes with the UL or CSA stamp.

OO - Provides a double layer of oil resistance with an oil resistant insulation.

SJO - Junior service cable with oil resistant jacket (300V if SJ, 600V if S).

SJOW - Junior service cable that is oil resistant and weather resistant (300V if SJ, 600V if S).

SJOOW - Full junior service cable that has a double layer of oil and water protection (300V if SJ, 600V if S).

These cords can be rated at light, medium, or heavy duty depending on the type of jacket on the cord.

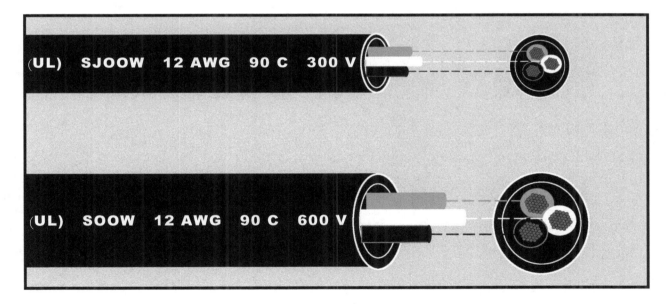

Power cords with thermoplastic jackets are light duty, those with thermoplastic elastomer jackets are medium duty, and cords with thermoset rubber jacket are considered heavy duty. Remember that **if the voltage is higher than 300 volts, then a power cord WITHOUT the J must be used.**

Extension Cords

For indoor use, extension cords are usually light duty and come in two-wire cords with either a white or brown jacket. Heavy duty cords are used in outdoor situations due to their weatherproofing and durability. They should also be used for high-wattage equipment or tools. These heavy duty orange cords are typically longer than indoor cords (25 to 100 feet).

UL approved cords will have the seal of approval on the female side of the cord. The mark may also be found as a mold on either the male or female end.

Ethernet Cables

Ethernet cables are commonly used in local area networks (LAN) to provide internet to both commercial and residential establishments. The cable design has changed several times over the last 20 years. We outline the major ones you will encounter below:

Cat5 cable:

Cat5 cables are an older generation of ethernet cords that are no longer installed but may currently be in place. They can handle up to 100 Mbps but have significant crosstalk, which is interference between the cords.

Cat5E Cable:

The Cat5E cable is a modified version of the Cat5 and has improved performance and speeds with reduced crosstalk. It can handle speeds more than 10 times greater than Cat5.

Cat6 Cable:

Cat6 cables are a more recent cable that improves upon the Cat5E cable. They were originally used as network cables and not for individual computers, but have since then become more popular. They can handle speeds up to 10 gigabytes of information without seeing decreased speeds, but this is limited to around 160 feet.

Cat6A Cable:

Cat6A cables are the most modern and advanced cables used today. They are similar to the Cat6 but have reduced crosstalk, and can maintain similar speeds at more than 300 feet. However, they can be expensive and unnecessary depending on their application.

3.2 Installing Cables

Learning to install cables overhead and underground is an important part of being an electrician. Also known as "pulling", we will discuss the methods of installing cables with framework in construction projects, as well as how to correctly bury outdoor electrical cables and wire.

Specific guidelines and requirements for installing cables can be found in the *NEC®* NFPA 70® manual. Always adhere to these rules when installing cables, and use the *NEC®* rules for box and wire requirements for each individual installation.

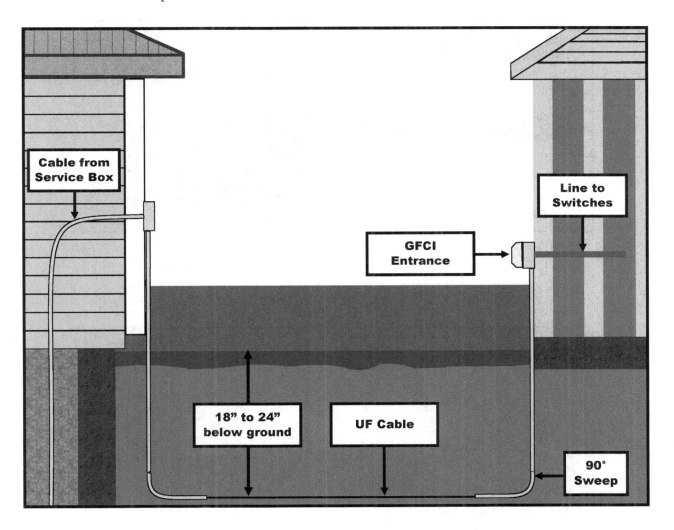

Running electrical power from the main household to additional buildings on the property can be done by using a service cable from the main panel. The underground cable should be run with approved underground (UF) cables at the appropriate depth.

Indoor Installations

In order to create a branch circuit from the main panel to be run throughout the structure, cables must be routed through the framework of the structure. This includes modification of the studs, rafters, or the floor and ceiling in order to properly pull cable.

Drilling Holes Through the Framework

You will need to drill holes in the framework in order to properly pull cables and secure them within the framework. You need to have the proper equipment, including UL approved power tools and the right size drill bits for drilling holes.

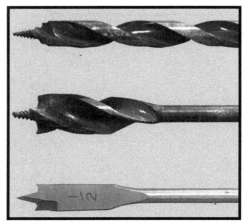 

Common bit types used for drilling holes in the framework usually range from 1/2 inch to 1 inch for electrical work. Use a power drill that is UL approved to place holes in the framework.

Holes in framework need to be at least 1 1/4 inch from the edge of the studs. If they are not, then a steel plate must be secured to the stud to prevent screws or nails from damaging the cable.

First Step of Circuit Runs

Each cable needs to be run into the SEP or subpanel box through the knockouts on the top, side, or bottom of the cabinet. Pull the required amount of cable through the knockouts to the point where there is more than enough cable line to connect the cables to the required circuit connections. Make sure the cables are secured with cable connectors so they cannot be pulled from the box during the remainder of the run.

Run a cable from the electrical cabinet to the required box. This is called a home run. You should **place at least 6 inches of cable in the outlet box for correct wiring and some extra for safe measure**. If you are unsure, you may feed excess cable into the box so that the cable may be cut to the perfect length at a later time.

***NEC®* article 334.30 states that once the cables leave the electrical cabinet or box they need to be secured within 12 inches from the knockouts.** This can be done with staples or cable clamps. Cables that are flat are required to be **secured on their flat side** for maximum safety as per *NEC®* code 334.30 guidelines. Once the cables are secured after leaving the electrical cabinet or box, **they should never be removed**. Only staples/ cable clamps after the initial mounting should be adjusted to prevent loosening of cables within the equipment.

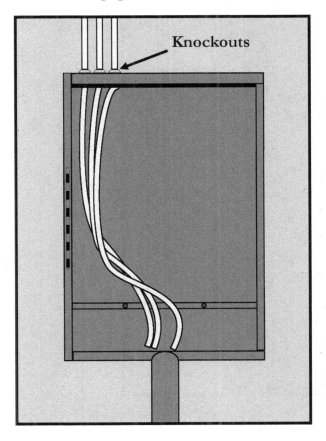

Knockouts

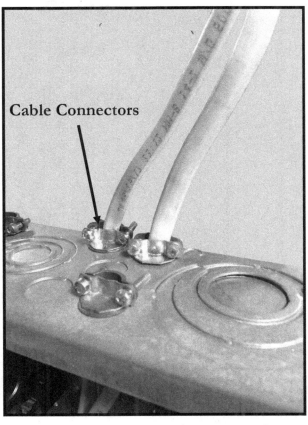

Cable Connectors

Cable Framework

To secure cable to the frame, electrical staples or cable straps may be used. Electrical staples are more commonly used because they are both effective and cost efficient. Ensure that the correct cable staples are being used, that they are fully inserted into the framework, and not bent or broken. The straps add an additional layer of security to the cable, but add an additional cost.

The type of box you use will also determine the incremental distance between each staple or strap. If you are using **nonmetal boxes, the cable must be secured within the first 8 inches** of leaving the box. If you are using a **metal box, then the cable must be secured within the first 12 inches** of leaving the box.

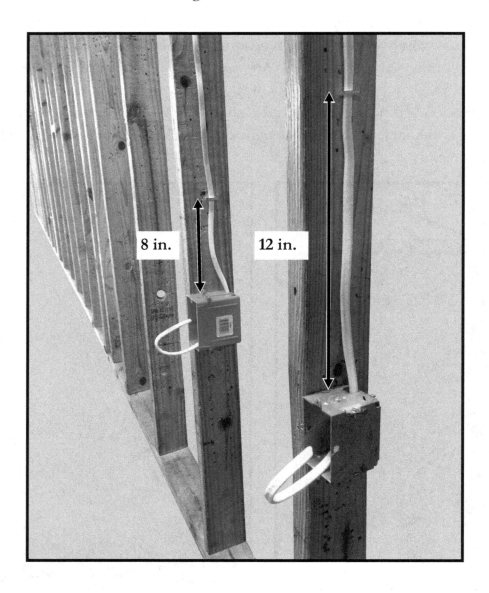

8 in.

12 in.

When you are running cable across studs or through them, the **cable must be secured every 4.5 feet for the entire circuit run**. When you run cable through a drilled hole in the stud, the distance between studs must be less than or equal to the 4.5 feet requirement. No additional securing measure is needed as **the holes support the cable**. In the case below, the cable is secured at 8 inches above the nonmetal box then another 4.5 feet above that, as set forth by *NEC®* code.

When performing a circuit run that is completed **using nonmetal boxes, the *NEC®* does not require any type of mounting of the cable to the box itself**. The *NEC®* does require that you leave the **cable sheath on the conductors for at least 1/4 inch after entering the box** to prevent damaging of the conductors from the knockout.

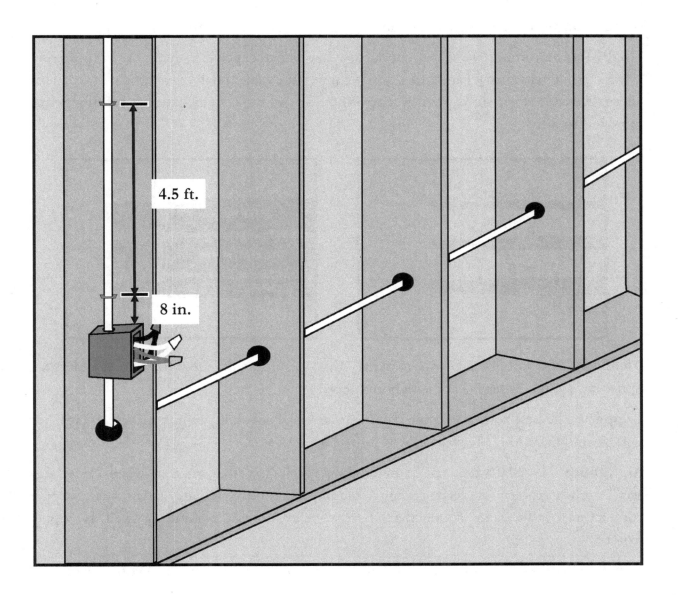

3.3 Electrical Conductors

Electrical conductors are the foundation for electrical work and movement of electricity. Electrical conductors allow current to flow freely through them by transferring electrons through the wire. When a voltage is applied to the conductor, these electrons start to move and allow current to flow freely.

An electrical "conductor" is any material that can freely move these electrons. Some conductors do this more easily, such as metals. **The best metal conductor that we know of is silver, but it can be expensive when compared to the more common copper**. Copper makes up the majority of wires utilized in the electrical field. Other cheap metals may conduct electricity, such as aluminum or iron, but are not as effective as copper.

Gold is also an important metal conductor. It is more expensive than copper and has a lower conductivity, but it can hold a steady conductive value much more efficiently than the alternatives. Gold is also resistant to corrosion, for this reason it is **used in plating when exposed to air.**

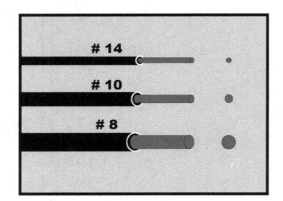 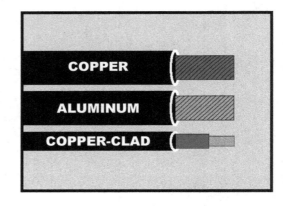

Aluminum and copper are two materials that are normally used for wire conductors. Sometimes you may have a combination of both.

Copper - The best conductor material due to its excellent ability to conduct electricity. It is durable and has a long life span.

Aluminum - Is used when you require larger conductor sizes, such as when using large service entry cables or feed circuits. Aluminum conductors are larger than copper conductors of the same conductivity because they do not conduct as well as copper material.

Copper Sheathed Aluminum - Aluminum core wire that is sheathed in a copper layer.

Electrical wires or conductors come in several different sizes and amperage ratings. Wire is sized based on the **American Wire Gauge (AWG) system**. This system gives **smaller number ratings to larger wire sizes, and vice versa**. When you start a new circuit run or rewire, make sure the conductor being used is rated to withstand the amperage provided by the circuit breaker in the electrical box. The wire size required will change as the amperage supplied by the circuit breaker increases. Larger diameter wires are needed to allow for the increased heat created by higher amperage circuits. The proper size is required to prevent electrical fires and reduce resistance.

When determining how much amperage is needed from the circuit breaker, you need to **consider the application of the wire. This includes any receptacles, lighting, or appliances** that will be powered through the conductor.

The difference in AWG sizing of wires comes with changes to amperage rating for each number designation. These **amperage ratings are based on the ampacity of each conductor**. This ampacity rating is based on the current a conductor can sustain without going over its given temperature rating. This guideline is provided by the *NEC®* to correctly classify AWG wire types.

AWG Wire Sizes	Amp Rating at 60° C	Amp Rating at 90° C
# 14	15	25
# 12	20	30
# 10	30	40
# 8	40	55
# 6	55	75
# 4	70	95
# 3	85	115
# 2	95	130
# 1	110	145

As the size of the wire increases, the need for stranded wire may arise. Stranded wires are different than single wire conductors. These stranded wires are normally used for large wires with #6 AWG or greater, but can also be found in smaller wires. **The cross-sectional area or circular mil of the stranded conductor is the same as a single solid conductor.** The stranded wire may seem larger than the same solid conductor, but its cross-sectional area is the same. For example:

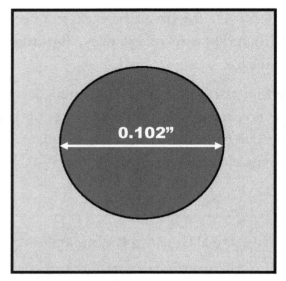

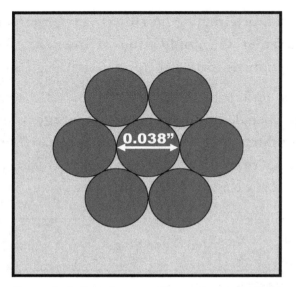

Single Conductor Cross-Sectional Area:

$$\frac{\pi}{4} D^2 = \frac{\pi}{4} (.102)^2 = 0.008\, in^2$$

Stranded Conductor Cross-Sectional Area:

$$\frac{\pi}{4} D^2 = \frac{\pi}{4} (.038)^2 =$$

$$.00114\, in^2 \times 7\, wires = 0.008\, in^2$$

In this example, we use an #10 AWG conductor which has a diameter of 0.102 inches. As you can see, the stranded wire appears to have more area at first, but actually has the same cross-sectional area as the single conductor. Because not all space between the stranded wires can be eliminated, it will be slightly larger than the same single conductor.

A circle "mil" is a unit we use when referring to the arc of a circle or in this case a conductor. One mil is equal to 1/1000th of an inch. You will often hear sizes described in mils on the jobsite. For this example with a **#10 AWG wire, the diameter in mils is 102 mils.**

While working you may encounter AWG wire sizes or amperage requirements different than those discussed in this section. For more detailed tables on temperature and amperage ratings of conductors, refer to **NFPA 70 *NEC®* Handbook** tables **310.15(B)(16) through 310.15(B)(21).**

Removing Cable and Wire Sheaths

Cables will need to be cut to expose the conductors. Take extra care to not damage the conductors when removing the cable sheath.

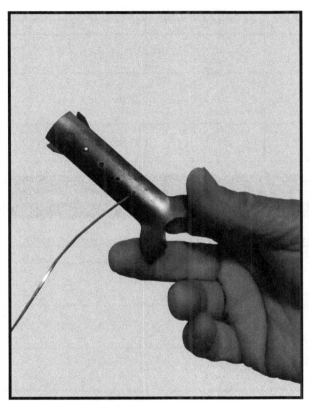

Most **cable strippers** have two number designations for each hole, the AWG and mm size. Cable strippers can remove the cable sheath by pulling the cable through the stripper.

A **wire stripper** is used to remove the protective sheath covering from the wire. Always work from the largest size first if you are unsure of the exact size of your wire. To strip the wire, insert it in the proper size gauge and close the device, pull the wire to strip the insulation off.

The tables below describe the **conductor requirements for simple circuit conductors** and their associated overcurrent protection for residential wiring. It also shows conductor requirements for **larger appliances which require their own branch circuit.** These should be used as a reference, but some commercial appliances may require different conductor sizes than the general ones listed here.

Conductor Sizing and Overcurrent Protection (Residential)		
Type	Conductor Sizing (Copper)	Max Overcurrent Protection (AMPS)
Dedicated	12 or larger	20 or larger
Small Appliance	12	20
General Purpose	14/12	15/20

Conductor Sizing and Overcurrent Protection (Appliance Branch Circuits)		
Type	Conductor Size (Copper)	Max Overcurrent Protection (AMPS)
Air Conditioner	12	20
Hot Water Heater	12/10	20/30
Oven	10	30
Stove-Top	10	30
Clothes Dryer	10	30
Central Cool/Heat	8/6	40/60
Self-Contained Range	6	50
Electrical Furnace	6 or Larger	50 or Larger

NEC® table 314.16(B) - Reprinted with permission from the National Electrical Code® Handbook 2017, copyright © 2017 National Fire Protection Association. All rights reserved

In addition, **two tables with cable nomenclature** can be found below. These letter designations can be found on all electrical wires and cables and should be chosen based on the installation location. **Refer to *NEC®* guidelines when determining what type of cable/wire is required.**

Cable and Wire Letter Designations	
F	Fixture Wire
FF	Flexible Fixture Wire
H	Heat Resistance (adds 15° C resistance to a 60° C conductor)
HH	High Heat Resistance (adds 30° C to a 60° C conductor)
HHW	Max operating temperature of 75° C when wet and 90° C when dry
N	Nylon outer jacket
R	Thermoset rubber insulation
T	Thermoplastic sheath rated at 60° C
W	Moisture (wet) Resistant
X	Thermoset cross-linked polyethylene insulation

Single Conductor Insulation Letter Designations				
Letter Mark	Insulation Type	Designated Use	Operation Temperature (Celsius)	Outer Jacket
TW	Thermoplastic (PVC)	Wet Environments	60° C	yes
THW	Thermoplastic (PVC)	Wet Environments	75° C	no
THWN	Thermoplastic (PVC)	Wet Environments	High Heat 75° C	yes (nylon)
THHN	Thermoplastic (PVC)	Wet Environments	High Heat 75° C	yes (nylon)

NEC® table 314.16(B) - Reprinted with permission from the National Electrical Code® Handbook 2017, copyright © 2017 National Fire Protection Association. All rights reserved

3.4 Splicing Conductors

When performing work, you will be required to connect wires together to add new devices or relocate the circuits. This is known as splicing. A splice joins two or more wires together by twisting them so that the exposed conductors carry an unbroken electrical current. It is very important to make sure these connections are secure and properly installed.

Loose electrical connections are a major cause of arcing and overheating, and can lead to electrical fires and injury. **To hold these connections in place, a plastic cap, called a wire nut**, is used both as an insulator and device for securing the wires together. Wire nuts come in various sizes specific to the number of wires being spliced and their sizes.

Each wire connector is made for a range of wires and sizes. As the gauge of the wire increases, the number of conductors the connector can hold will decrease. Each manufacturer will have unique standards for the wire nut they produce. Although a single brand of connector may be color coded, they are not the same colors for all producers. Therefore, you should always reference the packaging to determine the correct connector to use. If you do not have the package that the wire nuts came in, then check the manufacturer's website.

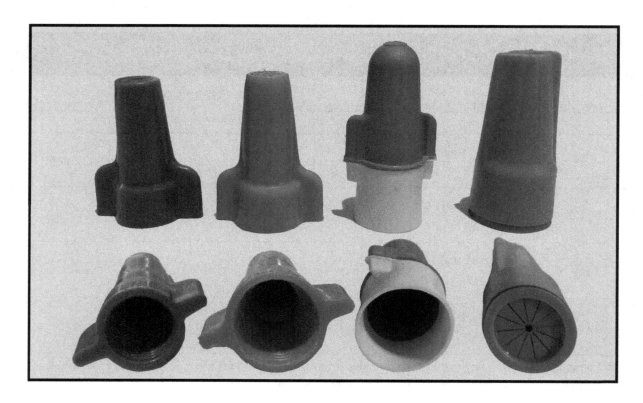

Solid and stranded conductors have different rules for splicing. When splicing solid conductors, the wires need to be lined up before they are twisted together and placed in the connector. These wires should be stripped before being connected. The strip length usually ranges from 1/2 to 5/8 inch, but you should always check the wire nuts recommendation on how much to strip.

As stated earlier in the chapter, stranded wires will be slightly larger than the solid conductor of the same gauge. To strip stranded wire, increase the hole size used to strip the wire by one. For example, when using #14 AWG wire, you should strip a stranded wire in a #12 hole.

Solid Conductor Connection

Make sure the exposed wires are lined up properly. Using lineman pliers, grip the two wires with enough pressure to twist the wires but not to break the conductors. Twist the two wires together in the clockwise direction for 1 to 2 complete turns.

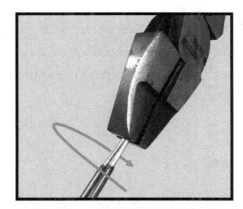

After the wires are twisted together, make sure the tips of the twisted wire are close together. If they are not, you may need to use wire snips to shorten one of the wires before attaching the wire nut.

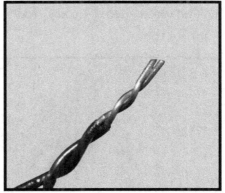

Attach the connector by twisting in the same direction, in most cases clockwise. The wire nut should be twisted until no bare wire is exposed. Slightly pull on the wire nut to ensure it is secure. If you exposed too much conductor when stripping the wire, electrical tape may be used to cover the wire and further secure the wire nut from removal.

Stranded Conductor Connection

Stranded wire must be connected differently than solid wire. **If you attempt to connect a stranded wire with a solid one the same way, the stranded wire will wrap loosely around the solid conductor and prevent the wire nut from securely locking the wires together.**

To make sure the stranded wire is twisted correctly, extend the end of the stranded wire by 1/8 inch past the solid wire. This will allow a secure connection when twisted together. The other steps to completing the connection are the same.

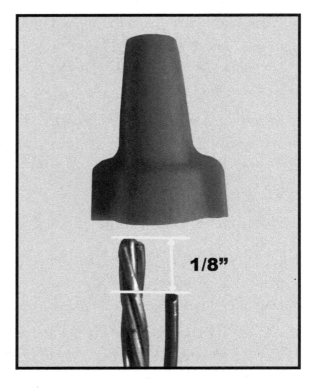

All spliced wire connections have to be contained inside a closed electrical box. These electrical boxes, called J-boxes, have to be sized correctly for the type of splicing being placed within them.

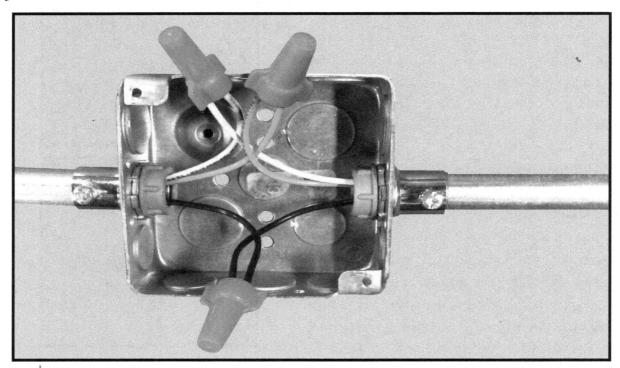

Other Connection Types

Cage Clamp®- easy to use and secure wire connectors that are installed by lifting the orange levers to open the lock, then inserting the wires and closing them. Make sure the connector is secure after installation. No tools are required to install these connectors which are classified as 3 and 5-pole connectors. They can accept #16 to #10 AWG wire.

Push Wire - are clear connection housings for splicing wires without the use of tools. They are self-locking, simply insert the wire to make the connection. Look through the device to ensure the wire has made the connection. They can be removed by twisting and pulling. They are produced in 2, 4, 6, and 8-pole with the ability to use #22 to #12 AWG wire.

Split Bolt - are used to splice two wires by clamping them together. This is useful when working with large appliances or trying to extend the length of a cable by splicing them together.

Insulated Power Connectors - these connections can be single-sided, double-sided, or spliced. Due to their insulation, these connectors do not need to be taped after installation. They are the most protected connection from moisture, chemicals, and abrasion. They can accommodate multiple conductors, either copper or aluminum, at 600V and 90° C.

3.5 Chapter 3 Review

1. A cable is a combination of multiple _____ sheathed together.

2. Two types of metal sheathed cables are _____ and _____.

3. The color of #12 NM cable is _____.

4. The color of # 10 NM cable is _____.

5. The abbreviation SER and SE stand for _____ and _____.

6. The letters S and SS on power cords stand for _____ and _____.

7. UF cable must be buried from _____ to _____ inches below ground.

8. Electrical cables need to be secured within _____ inches before entering an electrical panel.

9. #8 AWG size wire can carry _____ amps at 60° C.

10. Splicing two or more conductors together requires twisting wires and covering them with a plastic cover called _____.

CHAPTER 3 NOTES:

Working With Conduit

4

Electrical conduit is used to build pipe-way systems, also known as raceways. Conduit is used to protect the cables or wires held within. Using conduit will prevent moisture or impact damage, and will also increase the safety of the structure during construction. Conduit provides a path for power or low voltage applications. Similar to the different cables discussed earlier, conduit comes in a variety of sizes and materials. Each type and size has a specific use that will be discussed in this chapter. Conduit is almost always tubular, although some square conduit can be encountered when running wires outside of the wall.

There are also special types of conduit with added protection. When conduit needs to be placed in wet or dangerous areas, make sure it is rated to withstand the conditions it is placed in. The ***NEC® NFPA 70® Handbook provides specific guidelines for conduit installation and practices***. Using conduit has a number of advantages over simply using exposed cables, including:

1. more durable protection from environment or abuse

2. easier to pull cables and change wires through conduit vs a regular finished wall

3. can be sealed to prevent water or fire damage

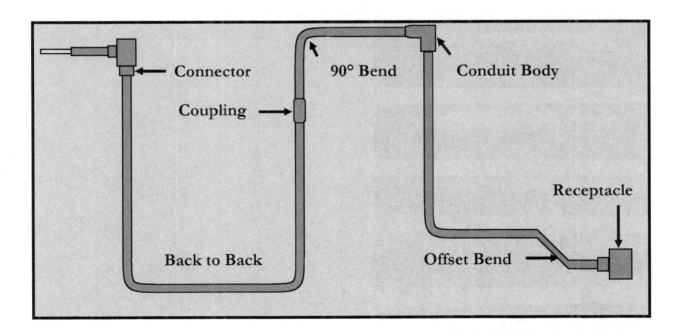

4.1 Types of Conduit

Electrical conduit can be either rigid or flexible and is classified in two main categories, metal and nonmetal types. Metal conduit is typically made of galvanized steel, stainless steel, or aluminum. Nonmetal conduit is typically made of a corrosive-resistant plastic. In this section we will discuss types of conduit, steps to installation, and their uses.

Different protective layers can be added to conduit by coating it or by galvanizing it. There are several different coatings that can be applied depending on where the conduit is going to be installed:

Galvanization - This process is used on RMC in commercial applications, and this RMC is often referred to as galvanized rigid conduit (GRC).

Bronze Alloy Coating - Used in highly corrosive environments such as refineries, chemical facilities, underwater, and on coastlines.

Aluminum PVC Coating - Used for RMC in corrosive environments where chemicals are present.

Rigid Steel PVC Coating - Used to resist acids, oils, and grease. This coating is also fire and water resistant.

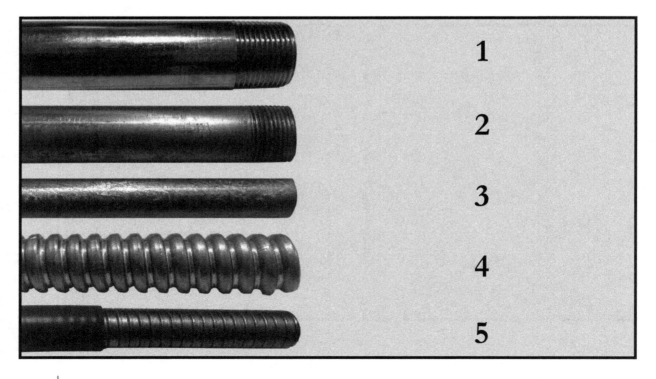

Metal Conduit

1. Rigid Metal Conduit (RMC)

RMC is a highly protective and sturdy conduit. It is thick and usually made of stainless steel or steel sealed in a protective coating. RMC may also be found in aluminum for some applications. RMC is threaded and each section can be screwed together.

As one of the most protective types of conduit, RMC prevents large impact and environmental damage. Because of its thick walls, RMC can be used to protect sensitive equipment from electromagnetic interference (EMI).

2. Intermediate Metal Conduit (IMC)

IMC has thinner walls than RMC, but thicker walls than other types of metal conduit. IMC is used in some residential applications and in the commercial industry. IMC is lighter than RMC, can be threaded, and also has coating options similar to RMC.

3. Electrical Metallic Tubing (EMT)

Another important type of rigid metal conduit is EMT. Electrical metal tubing is usually made of steel but can be made with aluminum when necessary. It is lighter than IMC and can be easily bent to fit a wide range of applications. The ability to form EMT in different directions makes it the ideal choice in many residential and industrial applications.

EMT does protect the cables that are run through it. However, EMT should not be installed in hazardous areas. If the conduit needs to be run in wet environments, or places where serious damage can occur (large plant installations), you should not use EMT.

4. Flexible Metal Conduit (FMC)

FMC is a nonrigid conduit that should be used in dry areas where it is not practical to use EMT. It usually comes in diameter sizes between 3/8" and 3". It is produced by coiling self-locking aluminum or steel strips. FMC is often used when protection of cables is necessary in moving equipment as it can reduce the vibrations passed from motor to structure.

You should always be able to distinguish the difference between FMC and metal clad (MC) or armored class cables that we discussed in the last chapter. FMC is a raceway to run cables through after it is installed, while MC and AC are sheathes used to cover wires and are permanent.

5. Liquid-tight Flexible Metal Conduit (LFMC)

LFMC is flexible metal conduit with a plastic coating and a waterproof seal. The conduit itself is similar to FMC but with the added benefit of being able to be used in wet conditions. It can also be buried and used to run cables underground. FMC coatings can come in various forms based on where the FMC needs to be installed. Coatings can be used when the FMC is required to withstand high temperatures, oils, corrosives, bacteria, or flames.

Conduit can also be galvanized rigid conduit (GRC). GRC is created by taking any of the rigid metal conduit talked about in this section and galvanizing it by dipping it in hot zinc. This coating protects the rigid conduit from corrosive chemicals and from rust due to oxidation or exposure to other liquids. For an additional layer of protection, PVC coating can be added to the GRC where there is a high chance of abrasion.

Nonmetal Conduit

EPVC

EPVC is a rigid nonmetal conduit that is the lightest and least expensive type of conduit. Sections of EPVC are fitted together using connectors and sealed with solvent. This process allows EPVC to be installed quickly. The solvent also makes the joints water tight. The most common types of EPVC encountered in the field are schedule 40 and schedule 80 pipe. It can be made with thicker walls if more protection is needed.

EPVC is resistant to moisture and corrosive substances, but is easily damaged from impacts when compared to metal conduit. EPVC can also expand or shrink with extreme changes in temperature.

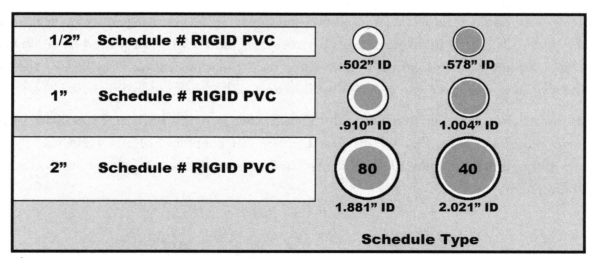

4.2 Bending Conduit

Learning to bend conduit for electrical installations is an important part of your profession. Being able to mold conduit to fit in residential or commercial wiring plans allows you to protect cables and wires no matter where the conduit is located. Remember, before you start to bend conduit you need to study the layout plan to ensure the angles you are using are correct. In addition, the **NEC® has specific guidelines for the number of angles permitted in lengths of conduit and should be referred to often.**

Conduit benders are used to form EMT in certain shapes to fit your wiring plan. Conduit benders come is various sizes depending on the size of the conduit you need to bend. Make sure to familiarize yourself with the specific instructions for the bender you are using, although most have universal techniques.

To guide you during the bending process, most conduit benders come with specific markings to guide your work. These symbols will provide you with angles and direction to properly bend each piece of conduit. Remember, you need to practice using these benders and making angles before working in the electrical field.

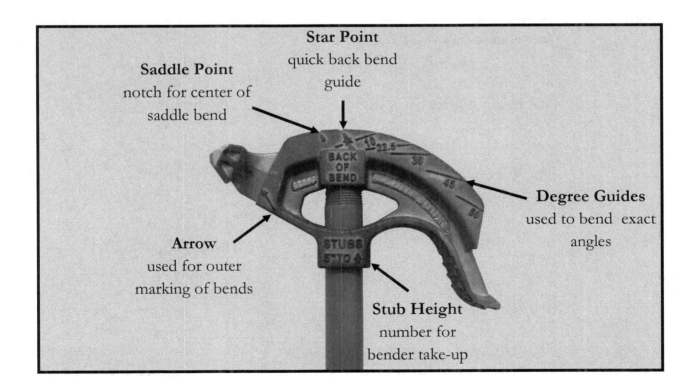

The 4 most common bends used are the 90° Stub-Up, Offset, Back-to-Back, and Three-Point Saddle bend. You will likely need to use several of these bends multiple times during the completion of one conduit run. There are a few guidelines you should follow when bending conduit:

- Use the correct size bender for the conduit.

- Properly measure and mark your conduit using bending tables before using the conduit bender.

- The bend should be rolled around the cradle in the bender. Always apply the bend with foot force.

- Depending on the type of conduit, there may be some spring back. Make sure to correct for this when bending.

- Remember to always secure the conduit during bending, never bend pipe if you are off balance or on an uneven surface. Make sure the bender will not slip when force is applied.

- Always wear gloves and safety glasses when bending conduit to prevent any accidental injury.

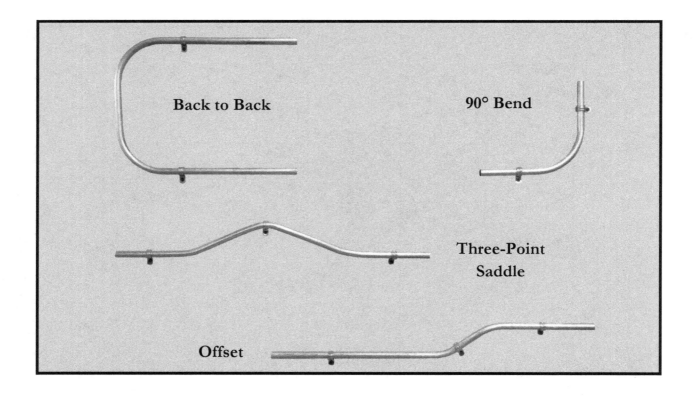

90° Stub-up Bend

This bend is the most common bend used for conduit and is used as a base for learning the other bends. 90° bends are used when turning up and down walls as well as running conduit into electrical boxes. The 90° bend makes a L-shape by placing the free end of the conduit into the pipe bender at a length according to the conduit size.

The following steps will guide you through performing a 90° stub-up bend:

1. Before you bend the conduit you need to determine the "free" end height you want. This height is how much conduit will be run after the bend on the free end.

2. Subtract the stub height based on the conduit size from the determined free height. The correct stub height can be found on the chart below.

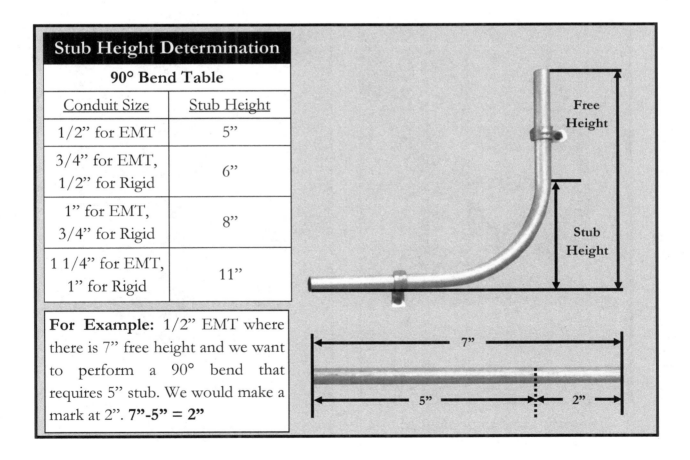

Stub Height Determination

90° Bend Table

Conduit Size	Stub Height
1/2" for EMT	5"
3/4" for EMT, 1/2" for Rigid	6"
1" for EMT, 3/4" for Rigid	8"
1 1/4" for EMT, 1" for Rigid	11"

For Example: 1/2" EMT where there is 7" free height and we want to perform a 90° bend that requires 5" stub. We would make a mark at 2". **7"-5" = 2"**

3. Using the correct size conduit bender, place the bender onto the free end of conduit. The hook on the bender should face the free end, or where your pipe is going. The hook should be placed under the conduit to be bent upward. Line up the arrow on the bender with the mark you placed on the conduit from the subtraction of the stub height.

4. Make sure the conduit is safely secure and flat on the surface it will be bent on. Apply the required foot force needed to the bender's heel. Do not apply excessive force to the handle, just add force on the heel where needed. Roll the conduit until the free end sits at 90° from the opposite end. There may be some spring back so you should measure with a level to ensure the angle is correct. If done correctly, the free end will be at the pre-determined height and the arrow will be located at the stub height.

Note: if you incorrectly measured the conduit and the free side of the conduit is too long, then you may need to cut it to install another length of conduit. This can easily be done using a hack saw or band saw.

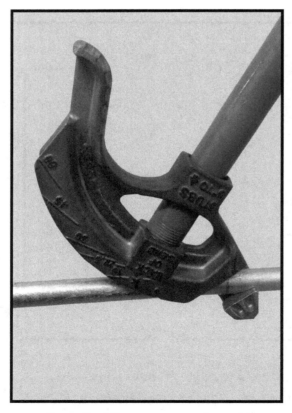

Step 3: Align arrow and mark

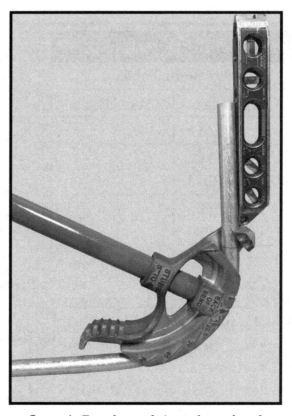

Step 4: Bend conduit and use level

Offset Bend

Sometimes you may need to change the position of the conduit while continuing to run conduit in the same direction. This can be done for a number of reasons, including obstacles or placement of electrical boxes. This bend is made with two separate bends.

The following steps will guide you through performing an offset bend:

1. Mark the offset distance necessary to clear an obstacle or reach an electrical box, and how far the offset needs to be bent from the end of the conduit.

2. Using the table below, determine the values needed based on the angle of the bend you want. Make sure to mark your conduit with the proper values before you bend.

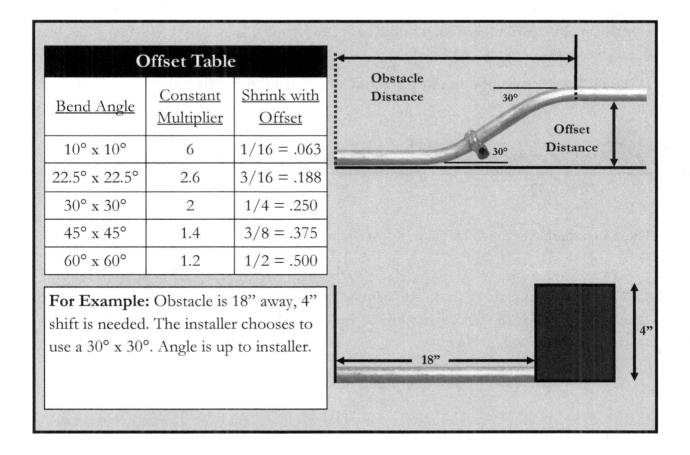

Offset Table		
Bend Angle	Constant Multiplier	Shrink with Offset
10° x 10°	6	1/16 = .063
22.5° x 22.5°	2.6	3/16 = .188
30° x 30°	2	1/4 = .250
45° x 45°	1.4	3/8 = .375
60° x 60°	1.2	1/2 = .500

For Example: Obstacle is 18" away, 4" shift is needed. The installer chooses to use a 30° x 30°. Angle is up to installer.

3. Using the table provided, follow the values given for 30° x 30° offset to make the right marks for your bend. To determine where to place the first mark on the conduit, multiply the offset distance to clear the obstruction by the value for shrinkage from the table. For this example:

(4") x (.250) = 1" total shrink

Add this value with the measured distance to the obstacle to determine the first mark:

(18") + (1") = 19" for first conduit mark

To determine where to place the second mark, multiply the measured offset distance by the constant multiplier number found in the table, this is the distance from the first mark that the second mark needs to be placed:

(4") x (2") = 8" from first mark, for a total of 27" from the start point

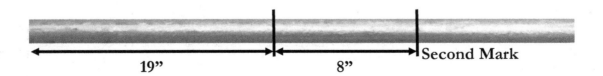

These two markings are where your 30° bends will take place.

4. To perform the first bend, you should place the bend tool similar to how you positioned it for the 90° bend with the hook facing away and the arrow lined up with the first mark. (See 90° bend)

5. Start to bend the conduit using the same technique of using foot force and pressure. For this bend, you will need to roll the bender until the 30° mark is reached on the bender. The free end of the conduit should be at a 30° angle from the floor.

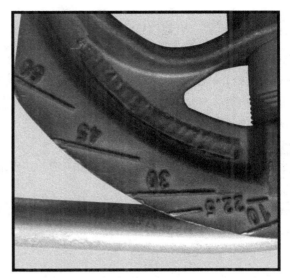

6. Making sure that the conduit and the bender stay together, flip both the conduit and bender upside down and place the handle of the bender on the ground with the conduit balanced on top. Rotate the conduit 180° and shift the conduit away from the bender head until the arrow lines up with the second mark.

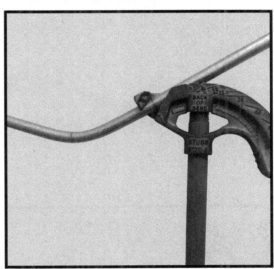

7. The second bend needed for the offset will be an air-bend. Make sure that the handle of the bender is secure on the ground, either with your foot or immobile surface. Make sure to correct your balance before starting the bend. Using the right amount of force with your body, bend the conduit around the cradle of the bender until the free end reaches the 30° marking. If you performed the bend correctly, then it will lie flat and clear the obstacle at the

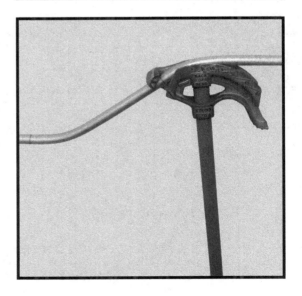

Back-to-Back Bend

Sometimes you will need to bend conduit between two walls or studs in order to switch directions or if you have no other option. The back-to-back bend creates a U-shape in the conduit. This U-shape conduit lays against each edge evenly so it can be securely anchored.

The following steps will guide you through performing a back-to-back bend:

1. Measure the distance between the two surfaces where the back-to-back bend will be placed.

2. The first bend you need to use is the 90° stub-up bend.

3. Based on the measurement from step 1, mark the conduit on the back end of the 90° stub-up bend.

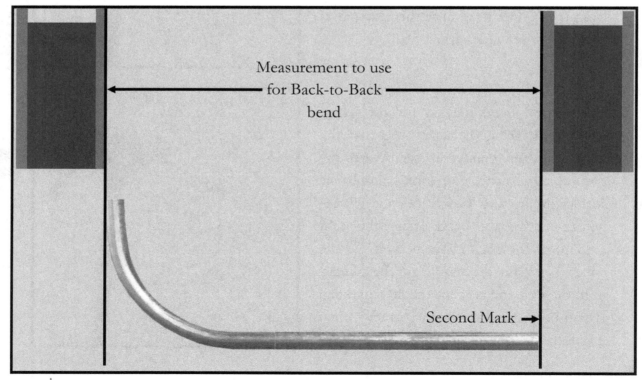

Measurement to use
for Back-to-Back
bend

Second Mark →

4. To prepare for the second bend, place the bender on the conduit with the hook facing the free end. The conduit will be bent to the opposite side of the first bend. For this bend, you will line up the second mark with the STAR symbol and not the arrow. This will keep the bend from exceeding the measured distance between the two surfaces.

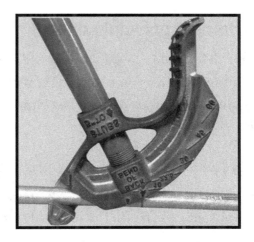

5. Secure the conduit then apply the correct techniques for a floor bend using foot pressure. Continue to roll the conduit until you reach the 90° mark.

Note: If the bender will not fit in the small area where the bend needs to take place, then you can subtract the stub height from the measured distance to give you more room to bend. If you do this, then you need to do the second bend from the arrow and not the star to compensate for the subtracted stub height.

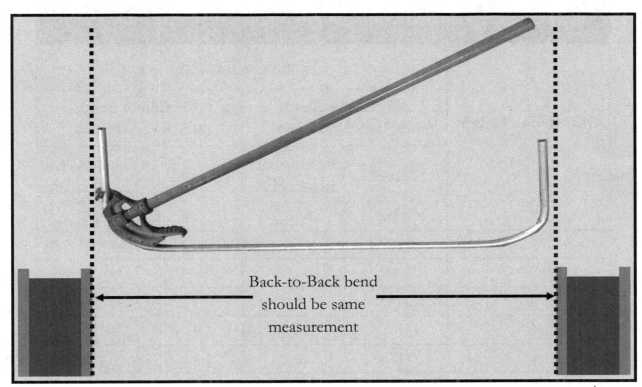

Back-to-Back bend should be same measurement

Three-Point Saddle Bend

This bend will be used when there is an obstacle in the way of the conduit but you want to continue in the same direction without an offset. This is normally used when there is existing conduit or other type of piping running across your path.

The following steps will guide you through performing a three-point saddle bend:

1. Measure the offset distance needed to avoid the obstacle as well as how far away the obstacle is from the end of the conduit. The distance to the obstacle should be measured to the center of that obstacle, so half of its total width.

2. Decide which bend to use for the center bend. The angle you use for the center bend will be cut in half for the other two bends. If you use a 45° bend then the first and third bend will be 22.5°. The table below will provide values for shrink and the distance needed between bends.

3. Determine the distance to the center mark on the conduit which is the measured distance to the central point on the obstacle in addition to the shrink from the table.

(Measured Distance to Center Point) + Shrink   22"+ 3/4" = 22 3/4"

	Three-Point Saddle Bend Table			
	Degree of the Bend			
Obstruction Height	**60° Center Bend** (30° for other bends)		**45° Center bend** (22.5° for other bends)	
	Shrink	Distance off Center Mark	Shrink	Distance off Center Mark
1"	3/16"	2-1/2"	1/4"	2"
2"	3/8"	5"	1/2"	4"
3"	9/16"	7-1/2"	3/4"	6"
4"	3/4"	10"	1"	8"
5"	15/16"	12-1/2"	1-1/4"	10"
6"	1-1/8"	15"	1-1/2"	12"

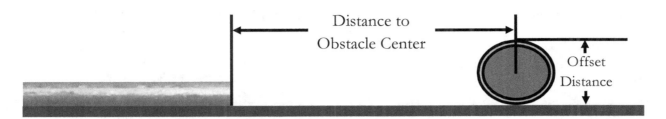

For example: the offset of an obstacle is 3" and the distance to the obstacle center is 22". We chose to use a 45° saddle bend for this installation. Remember that the choice of the angle for the bend is up to the installer but should be enough to clear the obstacle with the remaining conduit.

4. Use the table to determine the "distance off center mark" value for an obstruction with an offset height of 3". For the first mark, subtract this value from the distance to obstacle center. For the third mark, add this value to the distance to obstacle center.

First Mark:

(Center Mark) - (Distance off Center Mark)   **(22 3/4") - 6" = 16 3/4"**

Third Mark:

(Center Mark) + (Distance off Center Mark)   **(22 3/4") + 6" = 28 3/4"**

5. Mark the conduit with the first, center, and third marking.

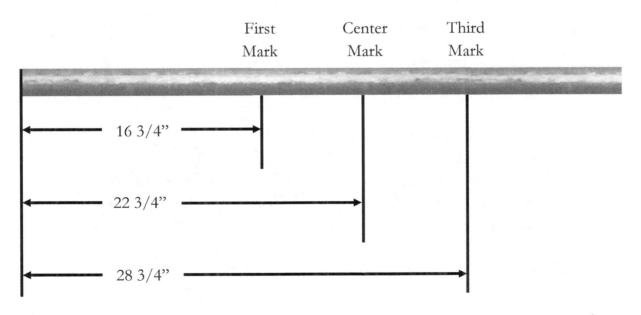

6. Place the bender on the conduit and find the tear drop notch on the bender. Place the bender with the hook towards the free end, align the tear drop notch with the center mark line.

7. Start to bend the conduit using the same technique of using foot force and pressure. For this bend, you will need to roll the bender until the 45° mark is reached. The free end of the conduit should be at a 45° angle from the floor.

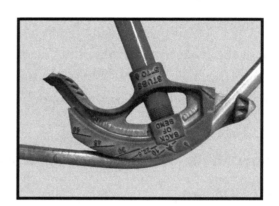

8. Making sure that the conduit and the bender stay together, flip both the conduit and bender upside down and place the handle of the bender on the ground with the conduit balanced on top. Rotate the conduit 180° and shift the conduit away from the bender head until the arrow lines up with the first mark.

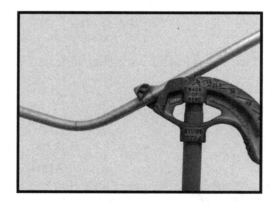

9. The second bend needed for the saddle will be an air bend. Using the right amount of force, use your body to bend the conduit around the cradle of the bender until the free end reaches the 22.5° mark.

10. Remove the bender and place it on the opposite side of the center mark. The hook should be facing the opposite way with the arrow lined up with the third mark.

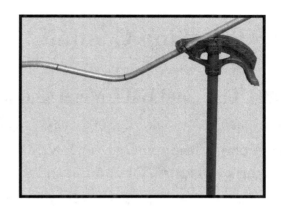

11. The last bend needed for the saddle will be another air bend. Make sure that the handle of the bender is secure on the ground, either with your foot or immobile surface. Using the right amount of force, use your body to bend the conduit around the cradle of the bender until the free end reaches the 22.5° mark.

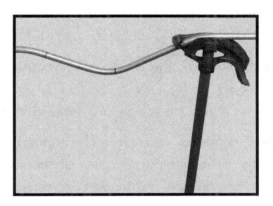

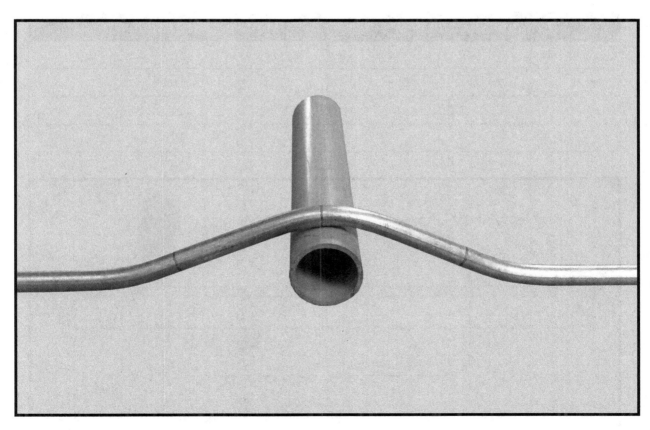

4.3 Installing Conduit

RMC, IMC, and EMT Metal Conduit

The steps to installing conduit vary greatly depending on where and what method of installation is being utilized. The **NEC® puts forth codes and requirements for the spacing of supports for conduit** and their frequency for RMC, IMC, and EMT.

These codes apply to conduit being installed in raceways, specifications for fastening, and distance between these points. These points, known as termination points, refer to outlets and junction boxes, cabinets, and conduit connectors or fittings.

For RMC, IMC, and EMT, **NEC® codes 342, 344, 346, and 358.30 provide guidelines.** The codes say that every 10 feet in the 1/2" to 3/4" conduit there must be an installed support. There must also be a support within 3 feet of device boxes, fittings, etc.

These supports are extremely important to reduce stress on the conduit and prevent damage or accidents. The size of the conduit also changes the required distance between supports. **NEC® table 344.30** shows how conduit size relates to the span between supports.

Conduit Size (inches)	Max Support Distance (feet)
3-6	20
2 - 2-1/2	16
1-1/4 - 1-1/2	14
1	12
1/2 - 3/4	10

Max 3 feet to box

3 ft.   6 ft.   6 ft.   3 ft.

Max 12 feet between supports for 1 inch rigid conduit

Supports for Nonflexible Metal Conduit Installation

For steel conduit, raceways can be mounted on the building's structure itself. The fasteners you use must be compatible with the conduit type and size you are using.

Installation guidelines for nonflexible metal conduit:

1. Steel conduit that is exposed on masonry, plaster, drywall, or wood surfaces may be secured using several fasteners that are approved for these surfaces. These fasteners include one or two-hole straps and conduit hangers.

2. Wood or metal framing can substitute as supports instead of fasteners when conduit that is run through openings in the frame. However, the termination fastening (3 feet within a box or device) must still be in place.

3. Steel conduit that is being run by suspension below ceilings or beams is recommended to be installed in pipe hangers. These pipe hangers are securely fastened with beam clamps and threaded pipe. You can also use a strut-type channel to provide support for your pipe.

4. If you are running raceways within a web of I-beams, then you may use column-mount supports for the conduit.

5. For a group of conduit running together, you should mount strut-type channels and straps to secure each group channel of conduit together.

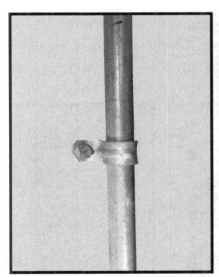  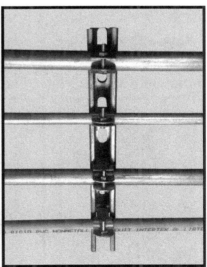

fasteners used for securing conduit on several types of surfaces

pipe hanger used to secure conduit to beams by clamps so it can be suspended

strut channels for securing multiple conduits which can vary in size

Fittings for Nonflexible Metal Conduit Installation

Connectors

Connectors join conduit to a device or box through the knockout. The connector's threaded end is inserted through the knockout and secured with a lock nut ring from inside the box. The other end of the connector can be secured to your conduit in several ways. For EMT, there is a set screw type or compression type connector. For RMC and IMC where there are no threads, a threadless type connector should be used.

Screw Type Connector

Threadless Type Connector

Compression Type Connector

Couplings

Couplings are used to connect two pieces of conduit together. Couplings can be used to create a conductive bond between pieces of the metal conduit. You may have to use grounding bushings which bond jumpers from the bushing to the grounding screw.

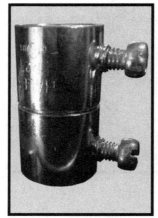

Screw Type Coupling

Threadless Type Coupling

Compression Type Coupling

Threaded Type Coupling

Threadless fittings cannot be used with threaded conduit such as RMC or IMC unless the fittings manufacturer specifically states it is made to do so. When you are to install threadless fittings to RMC, IMC, or EMT, the conduit ends should meet the following specifications:

- Be clean from dirt or debris on the surface of the conduit being inserted into the fitting.

- The conduit should have squarely cut ends and be free of internal or external burrs. It should also have a circular form as produced by the manufacturer.

- Make sure the conduit fits securely against the inside edge of the fitting and that the conduit is inserted to the stop point in the fitting.

Threadless fittings used in wet locations will have special markings that indicate their special use. These markings are either found on the fitting itself or on the shipping box. The designation of "wet locations" may also be stated as "raintight" on the fitting.

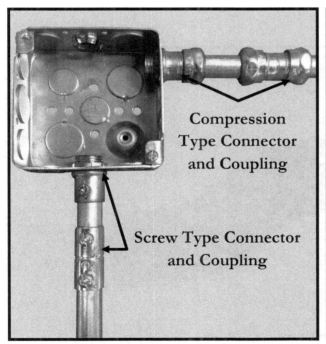

Compression Type Connector and Coupling

Screw Type Connector and Coupling

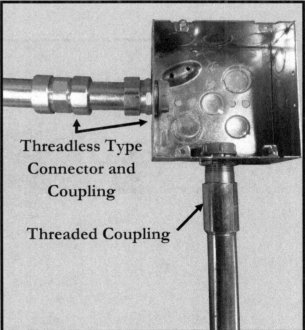

Threadless Type Connector and Coupling

Threaded Coupling

Expansion and Deflection Fittings

We use expansion and deflection fittings to combat the process of expansion or shrinkage of the conduit due to temperature changes. Expansion fittings are required to allow the conduit to adjust in the fitting depending on the temperature. They should always be used, even where there is little temperature change. **A change in temperature as little as 20° F can cause the conduit length to change by more than 1 inch.**

Deflection fittings are seldom used and are a relatively new fitting used in the electrical industry. They are used to slightly angle around obstructions without the use of bending conduit or using angle joints. They can also be used when two sets of conduit lines are being connected and are at slightly different heights or angles.

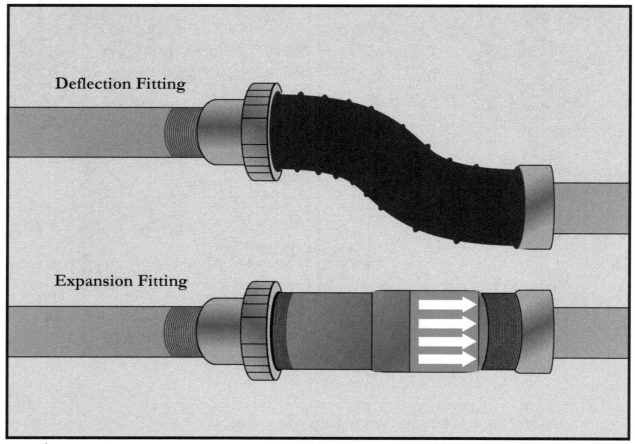

Deflection Fitting

Expansion Fitting

Conduit Bodies for Nonflexible Metal Conduit Installation

Conduit bodies allow you to make more turns and bends in the conduit. They provide pulling space when you need to make sharp turns where normal conduit bends will not work. Conduit bodies can also be used to split the raceway into separate paths.

Conduit bodies are not required to be supported themselves. This makes them convenient for raceways. Conduit bodies are often referred to as condulets and they come with various ratings and materials similar to conduit.

Conduit bodies are classified under several types depending on their direction and the material they are made of. These bodies, depending on the material they are made of, can be secured differently.

L-shaped bodies (Ells) - The inlet is in line with the access cover with the outlet placed on the back on either side. They can be LB, LR, or LL bodies.

C-shaped bodies (Cees) - Are used to pull wire through a straight run.

T-shaped bodies (Tees) - Inlet is in line with the cover with two outlets on either side of the inlet.

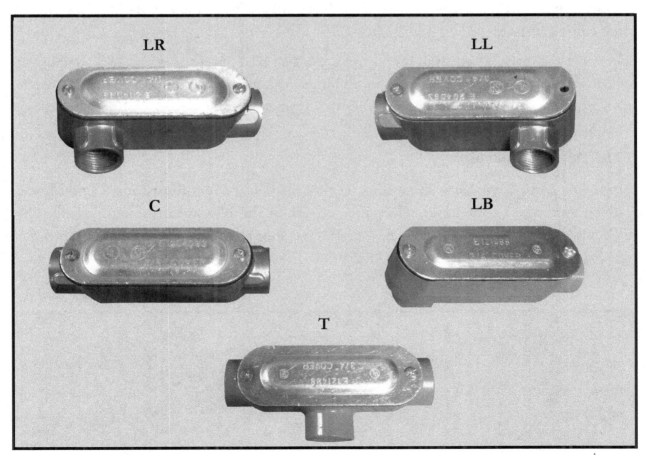

Flexible Metal Conduit (FMC)

The fastening requirements for flexible metal conduit are different from other types of metal conduit. **Sections 348.30(A) and (B) of the *NEC®*** guidelines provide requirements for proper installation of flexible metal conduit. In general, FMC must be secured within 12 inches of an enclosed device, box, etc. Intervals for securing FMC past the initial fastener can be found in the table below.

When flexible metal conduit is used, the length measured from the last point in the raceway where it was secured must not exceed the following values:

Flexible Conduit Size (inches)	Max Support Distance (feet)
2 1/2 +	5
1 1/2 - 2	4
1/2 - 1 1/4	3

Cutting FMC

Measure your FMC and align the cut mark. Make sure to place the cut mark off the end of your cutting surface.

There are two ways to cut FMC. You can use a fine blade hacksaw to make a straight cut with little to no angle, or you can use a cutting tool which may be faster but can create dangerous sharp angles. You should smooth any angles or sharp edges before connecting the FMC to your device or box.

FMC Device Installation

Once your FMC is cut, place a bushing on the side that was cut to provide a smooth and safe connection to your fitting. Place the cut side of the FMC with the bushing into the lock nut FMC connector and tighten the screw to secure it. Place the fitting assembly into the device box and secure it with the lock nut.

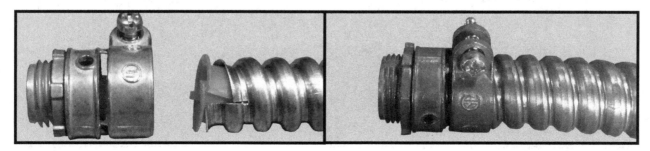

PVC Conduit Installation

When installing PVC conduit you will normally be installing either schedule 40 or schedule 80 PVC. Both of these types have the same outer diameter, however the inner wall of the schedule 40 is smaller than the 80, which means it is a thinner PVC. This makes it easier to run cables through, but it is not as sturdy as schedule 80. Schedule 80 PVC conduit should be used where there is any chance of it being damaged or if it is in a high traffic area.

Cutting PVC conduit

There are several ways to cut PVC conduit due to its plastic material. The easiest and most efficient way is with a circular saw using a metal blade. This will leave the cut end of the PVC conduit smooth without burns. If you do not have a metal blade or circular saw, a regular cutting saw or blade will do.

Assembly of PVC conduit and fittings

The attachment of fittings on PVC conduit is fast and easy if done properly. The joints will be both air and water tight. You should always deburr conduit if it is cut before being glued. Make sure the conduit is also clean. The following are steps to connect PVC to fittings:

1. Apply the PVC primer to the outside of the conduit and the inside of the fitting. Apply a small layer of the primer to the surface making sure to cover all the area that will be touching for the connection.

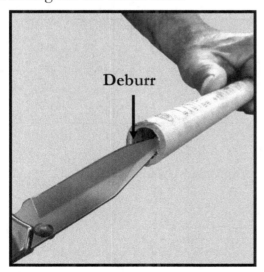

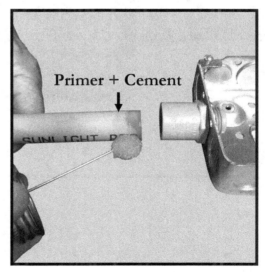

2. After using the primer you should apply the PVC cement the same way. Once the cement has been applied you should connect immediately.

3. Insert the PVC conduit into the fitting and push until the conduit can go no further inside the fitting. When the conduit can be inserted no more, apply a 1/4 turn twist to secure the connection and hold it in place for 15 to 30 seconds.

4. Make sure to allow the bond between the fitting and the conduit to dry and harden for at least 30 minutes before applying any force to it.

Using primer - Primer is normally used by plumbers when installing PVC plumbing to give extra protection against water leakage. In the electrical field, it is acceptable to use primer but many electricians will skip this step unless it is required by the conduit maker or building codes.

Securing and support of PVC conduit

Like metal conduit, PVC conduit must be secured within 3 feet of entering or leaving any type of box. However, the codes for securing EPVC after the first fastener are different than metal conduit. Each support distance will increase as the size of the conduit gets larger. **Refer to *NEC®* table 352.30 for information on sizing and fastening lengths.**

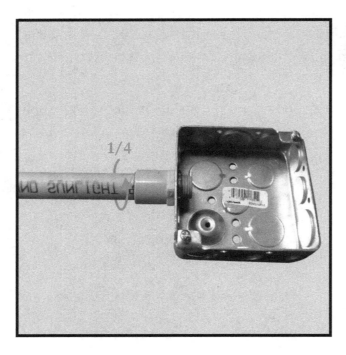

Conduit Size (inches)	Max Support Distance (feet)
6	8
3 1/2 - 5	7
2 1/2 - 3	6
1 1/4 - 2	5
1/2 - 1	3

PVC conduit is more susceptible to shrinkage and expansion due to temperature changes than other types of conduit. When the PVC conduit gets hot it will expand, and when the temperature drops it will shrink. The combination of these events may cause the EPVC to disconnect from fittings, straps, or boxes. Knowing how much a length of EPVC will expand due to temperature changes throughout the year can help you avoid having to deal with separated PVC lines. **Refer to *NEC®* table 352.44 for details about temperature changes as well as expansion and shrink lengths.**

PVC fittings and connectors

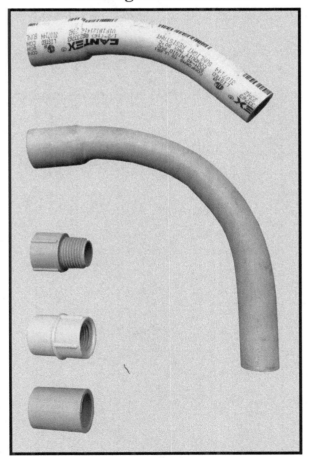

PVC expansion fitting

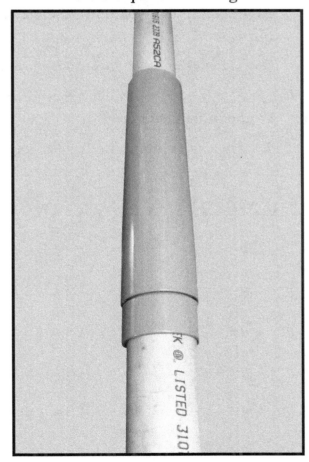

4.4 Chapter 4 Review

1. Three types of conduit are _____, _____, and _____.

2. EMT stands for _____.

3. The star symbol on a conduit bender denotes the _____ of a bend.

4. For a 1/2" EMT stub up bend, you should deduct _____ inches.

5. The multiplier for a 30° offset bend is _____.

6. The second bend performed on a 3-point saddle bend will be aligned on the _____ symbol located on the bender.

7. Rigid conduit needs to be supported within _____ feet of a panel or fitting.

8. When installing EMT outdoors you must use _____ fittings.

9. _____ must be used to allow for length changes of conduit due to temperature changes.

10. Two types of conduit bodies are _____ and _____.

CHAPTER 4 NOTES:

Electrical Circuits

5

Electrical circuits are the foundation for the transfer of electrical power. Electrical circuits are paths which allow electrons to flow from a power source. When these electrons leave the electrical circuit path it is called a return and travels along a path known as an earth ground. The term "return" is used because the electrons in the circuit always return to the source to complete the electrical path.

The point between the source and the return on the electrical circuit is called the load. These loads are created by appliances or fixtures which have different load ratings.

<u>In this chapter we will discuss the following topics related to circuits:</u>

1. Electrical Circuit Principles

2. Types of Electrical Circuits

3. Feeder Circuits

4. Branch Circuits

5. Planning Circuit Locations

5.1 Electrical Circuit Principles

When current enters a residence or commercial structure it is carried through an ungrounded (hot) circuit conductor and is said to be "under pressure". This voltage under pressure moves current through the residence to power lights, receptacles, and appliances. When **it is finally returned to the power source through the grounded (neutral) circuit conductor** it is considered under zero pressure.

A black conductor is designated to be the ungrounded circuit conductor which carries the voltage from the power source to the loads in the structure. The white conductor is typically used as the grounded circuit conductor (neutral) which is used as a return to the power source under zero voltage.

When installing receptacles or switches, an additional copper wire, that is not sheathed, can be used to ground the device and provide an extra layer of protection for people working with the circuits.

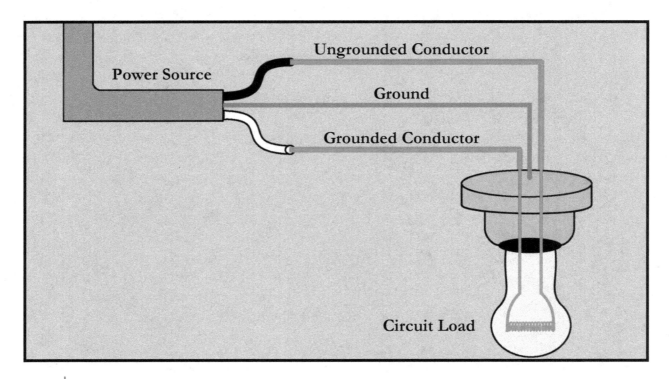

5.2 Types of Electrical Circuits

There are two types of electrical circuit designs that you will encounter when wiring:

1. **Feeder Circuit** - Feeder circuits run from a piece of service equipment to a separate system or some other power supply. Unlike the branch circuit, the feeder circuit runs to the final branch overcurrent device instead of from it.

2. **Branch Circuit** - The branch circuit is designated by the *NEC®* as "the conductors between the branch circuit final overcurrent device protecting the circuit and the outlets". Branch circuits typically run from panels to devices which hold load. When you compare the installed branch circuits and feeder circuits in the same residence, the branch circuits will have lower voltage than the feeder circuits.

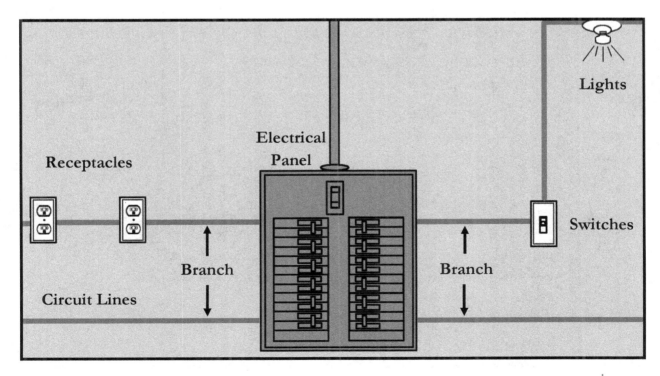

5.3 Feeder Circuits

Feeder circuits are used to supply power to subpanels from the service entrance panel. Feeder circuits are also additional power supply sources such as batteries or generators. As a structure increases in size, the addition of subpanels makes it easier to wire locations that are far from the service entrance panel. Using subpanels allows for an easier wiring plan and also adds a safety barrier to unexpected voltage drops from using long distance branch circuits.

In the past few years, the *NEC®* has created specific guidelines for the safety of wiring for service panel and subpanel installations. **New installations in recently built structures are required to have an electrical power disconnect outside the area of the structure.**

This is usually accomplished by placing the service entrance panel with a disconnect outside the structure with the subpanel being placed inside the residence. These *NEC®* codes have become the standard for residential wiring over the last few years.

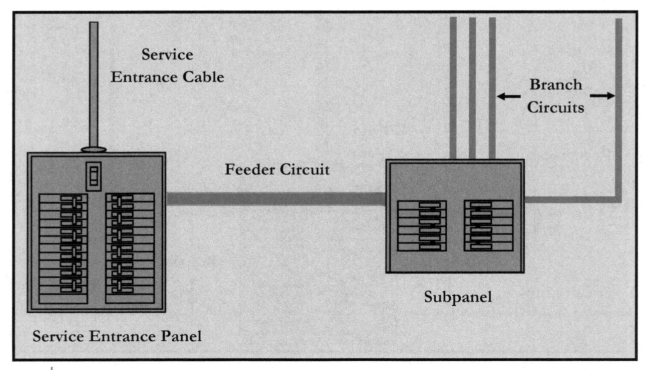

5.4 Branch Circuits

There are several types of branch circuits that you will encounter when wiring residential or commercial structures. These types can be described in 4 categories:

1. General Purpose Branch Circuits

2. Small Appliance Branch Circuits

3. Individual or Dedicated Branch Circuits

4. Bathroom Branch Circuits

General Purpose Branch Circuits

These branch circuits are 120V and are primarily used to power small fixtures and receptacles for small appliances. General purpose branch circuits are a very common installation type and are normally the most used type of branch circuit in residential and commercial branch installations.

The number of branch circuits in any installation is determined by multiplying 3 volt-amperes per square foot by the dimensions of the home and then dividing by 120V times the amperes. This will give the number of 15 or 20 ampere light area circuits.

[length x width x 3 / (120 x [15/20])] = number of 15/20 ampere circuits allowed

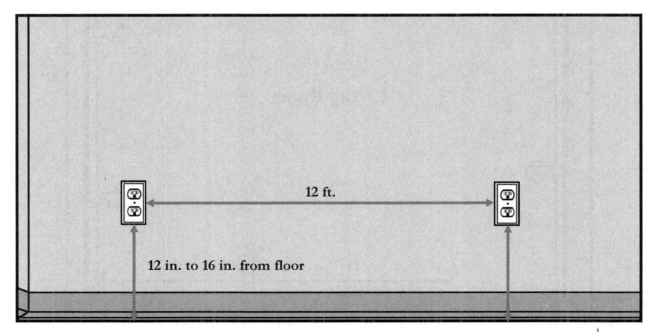

There is an easier way to determine the number of branch circuits. When counting the number of devices connected to each receptacle, you can assume that the outlet has 1.5 amps. The *NEC®* states that you can only take 80% of the rating of the outlet. Therefore, the total load on each outlet (all the devices connected) cannot exceed a 12 amp load on a 15 amp breaker circuit and cannot exceed a 16 amp load on a 20 amp breaker circuit.

General Purpose Receptacles

There are no height requirements for outlet installation put forth by the *NEC®*. However, the accepted industry standard is for outlets to be 12 to 16 inches from the ground. The *NEC®* does require that outlets on an unbroken wall must be within 12 feet of each other. Wall space of 2 feet or more that is not broken by doorways or openings requires a receptacle. This requirements applies to all rooms in the wiring installation. If walls are separated by doors or openings then these walls are to be counted separate from each other. **For more references on outlet spacing, refer to *NEC®* article 210.52 and 210.26.**

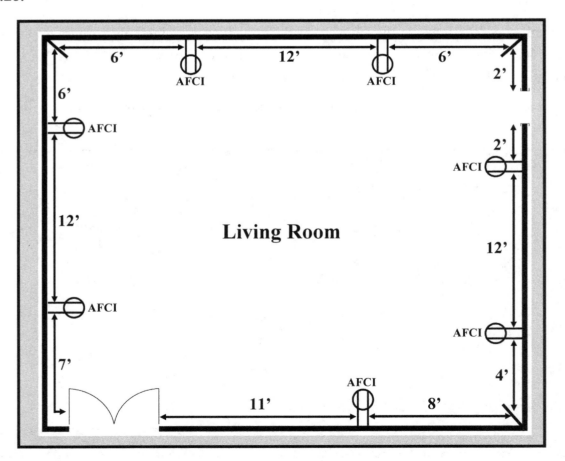

Dedicated Branch Circuits

Dedicated branch circuits are different from other circuits in that they require their own circuits when connected to the breaker. These are normally needed when installing large appliances. These can include but are not limited to:

- Refrigerator/Freezer

- Dryer

- Microwave

- Dishwasher

- Garbage Disposal

- Range

- Heating/Cooling System

These large appliances usually require anywhere from 20 to 50 amperes to run effectively. If these appliances are not wired correctly or given enough amperes from the breaker, they may not function properly or at all. In addition, overloading may take place.

Overloading occurs when the wire size or breaker are not appropriate for the appliance installation. This can cause overheating of the wire which can start an electrical fire. There are safety mechanisms to prevent this, such as tripping the breaker, but this should not happen if proper wiring protocol is used.

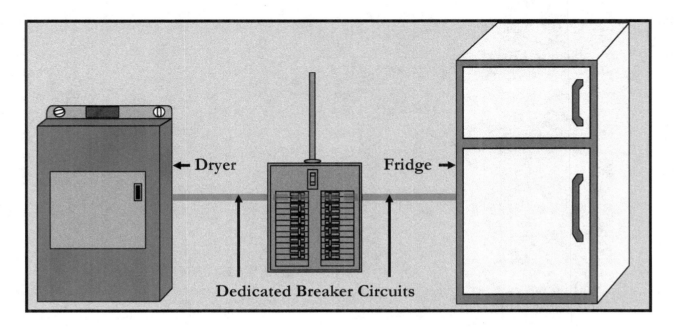

Dedicated Range and Dryer Hardwire

The *NEC®* allows some dedicated circuits to be hardwired without the use of receptacles. These connections are made between the terminal blocks located on the range and dryer to the overcurrent protection device in the electrical panel. There are two types of hardwire installations depending on if the cable being used is a new 4-conductor cable or a 3-conductor cable that is already installed.

Range - requires a #6 AWG wire in a 3-conductor cable + ground and an overcurrent protection device of at least 50-ampere.

Clothes Dryer - requires a #10 AWG wire in a 3-conductor cable + ground and an overcurrent protection device of at least 30-ampere.

Using a 4-Conductor Cable for New Installations

- Connect the red (hot) and black (hot) conductors to the brass terminals on the block.

- Connect the white conductor to the silver terminal on the block.

- Connect the grounding conductor to the green grounding terminal on the block.

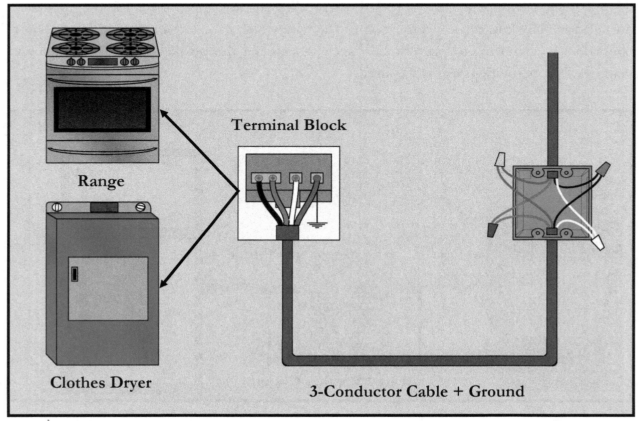

Using an Existing 3-Conductor Cable for Installations

Three conductor cable installations are an outdated version of the range stove and clothes dryer hardwire installations. **The *NEC®* stopped allowing 3-conductor cables for hardwire installations in the 1990's and switched to the 4-conductor cable.** However, you will often encounter older installations when doing maintenance or replacing existing hardwired appliances. Therefore, it is important for you to be familiar with these types of installations.

The terminal blocks on appliances are the same for the 3-conductor cable or 4-conductor cable installation. No matter what type of installation you are performing, remember to adhere to the appliance manufacturer's guidelines as well as *NEC®* codes and regulations.

- Connect the red (hot) and black (hot) conductors to the brass terminals on the block.

- Connect the white conductor to the silver terminal on the block.

- Connect a green jumper wire from the silver terminal to the green grounding terminal. In older installations the green grounding jumper is pre installed by the appliance manufacturer. However, it may need to be bonded to the block and installed manually for new hardwire installations with older 3-conductor cables.

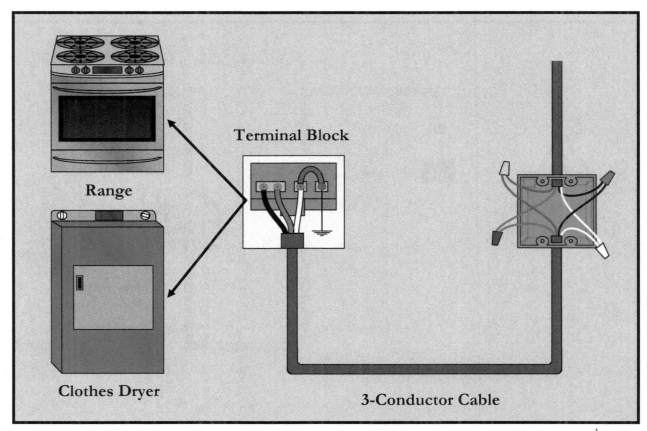

Range

Terminal Block

Clothes Dryer

3-Conductor Cable

Dedicated Water Heater Hardwire

Water heater installations use a 2-conductor + ground cable for connection to the overcurrent protection device on the electrical panel. Because water heaters have a high potential for moisture damage to the cable, there are specific guidelines for distances between the hardwire connections and the location of the water heater. Most codes require at least 1 foot from the hardwire connection to the water hater, and some local codes may require more distance.

In most cases, the water heater will not be located in the same room as the electrical panel, as in water heater attic installations. In this case, the *NEC®* requires that a disconnect be used within 50 feet of the water heater that can be used to cut the connection between the water heater and the electrical panel.

Water heaters function at 240V and require a 30-amp overcurrent protection device. This usually means #10 AWG wires are required. If the water heater is smaller and runs at 120V, then #12 AWG wires can be used with a overcurrent device protection rated at 20 amperes. Determine which wire size you should use based on local codes and the manufacturer's guidelines for the water heater.

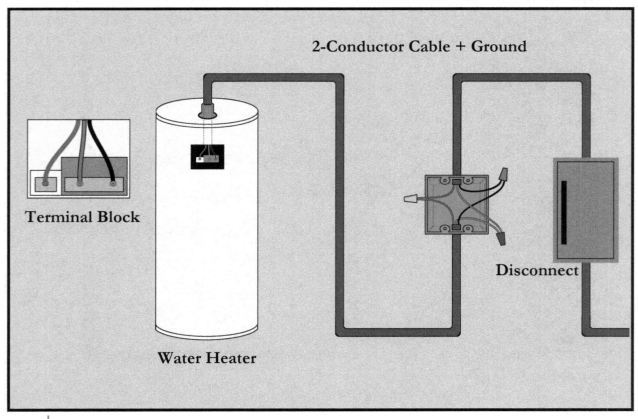

2-Conductor Cable + Ground

Terminal Block

Disconnect

Water Heater

Separate Oven and Cooktop Hardwire

Using hardwire connections for a separate cooktop and oven is very similar to a single connection. The conductors from the electrical panel, the cooktop, and the oven are all spliced together in a junction box. **The three cables used in the hardwire installation should be 4-conductor cables (3-conductor + ground).**

The cables run to the cooktop and oven may be NM cables or conductors housed in flexible metal conduit (FMC), which is usually provided by the appliance manufacturer. If you choose to use the FMC installation, then a metal junction box must be used for housing the connections.

The required overcurrent protection and AWG size will depend on the manufacturer's requirements for the oven and cooktop. These requirements along with regulations put forth by the **NEC® on table 8-1 of the NFPA 70®** will guide you in proper overcurrent protection requirements for these hardwire installations.

The terminal blocks for each of these installations will be the same as the single range terminal block. Remember, in older installations you may encounter the green jumper wire installation, in this case you should follow the same guideline as the older range installation.

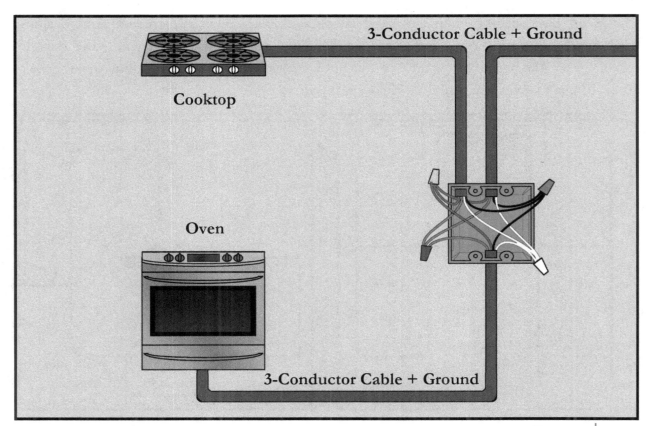

Small Appliance Branch Circuits

Small appliance branch circuits are used to power devices on kitchen countertop surfaces and small appliances in the dining room areas. There are a range of appliances in the kitchen that can utilize these circuits, including coffee pots, toasters, electric grills, etc.

Only GFCI/AFCI outlets are allowed to be used for small appliance circuits on the kitchen countertop. However, regular small appliance circuits can be used to power the electric ignition for ranges.

Smaller refrigerators may also be able to use regular small appliance circuits.

Kitchen countertop receptacles cannot be placed more than 20 inches above the counter surface.

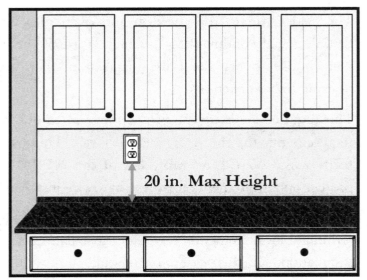

20 in. Max Height

There are also specific guidelines for how small appliance circuits are run around sinks. Below on the left you can see that receptacles are not required when a sink is less than 12 inches from the wall. The picture on the right shows that if a corner mounted sink is less than 18 inches from the wall then it does not require outlets behind the sink.

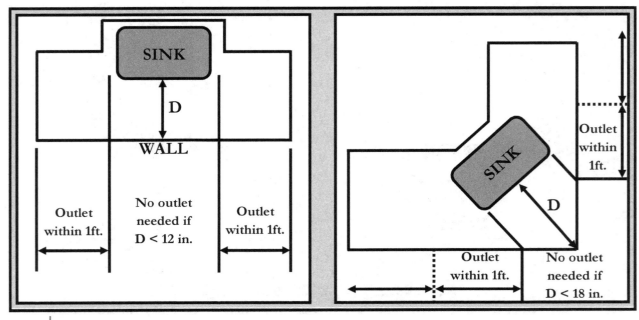

The *NEC®* sets guidelines on the requirements for installation of small appliance branch circuits. In addition to the spacing requirements listed below, these branch circuits must use #12-2 AWG wire or smaller with a ground. In addition, each circuit must be protected with a 20-amp GFCI/AFCI circuit breaker.

Small Appliance Receptacles

The *NEC®* also sets guidelines requiring that the distance between outlets on kitchen countertops cannot exceed 48 inches. If the counter space is more 12 inches long, then at least one receptacle is required for that area. The floorplan below shows the spacing of receptacles on the kitchen counter.

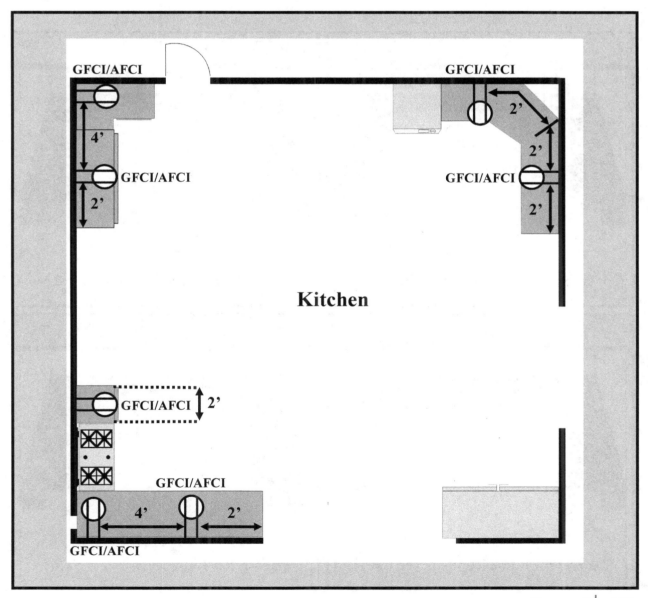

Small Closet Light Devices

Lighting in closet space is important and the *NEC®* has put forth specific guidelines for spacing requirements. There are two sets of guidelines for spacing depending on the type of light and how it is mounted to the ceiling. For **incandescent lighting**, the storage shelves should be 12 inches or more from the light fixture. If these lights are recessed, then the requirement is 6 inches or more to the actual light housing.

Fluorescent lighting that is surface mounted requires at least 6 inches of space to the end of the fixture. When it is recessed, it also requires at least 6 inches to the bulb housing, similar to the incandescent recessed lighting.

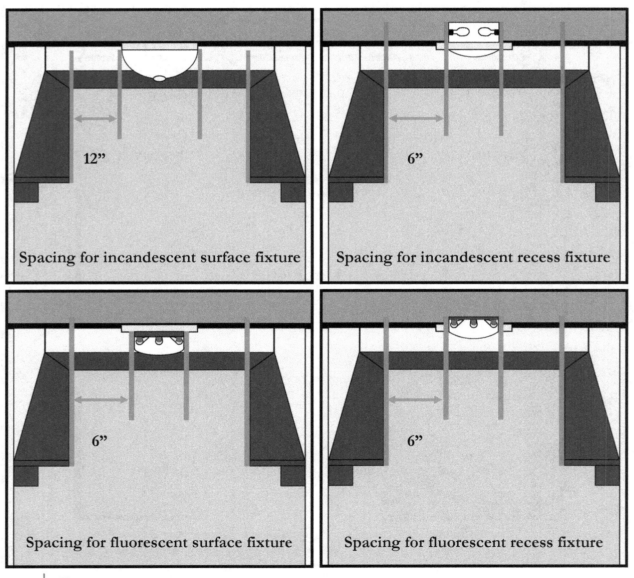

12"	6"
Spacing for incandescent surface fixture	Spacing for incandescent recess fixture
6"	6"
Spacing for fluorescent surface fixture	Spacing for fluorescent recess fixture

Bathroom Branch Circuits

Bathroom branch circuits have guidelines that differ from all other types of branch circuits. **All outlets located within the bathroom must be GFCI protected**. GFCI receptacles protect against the raised possibility of plugs coming into contact with liquids.

The **NEC®** has also recently changed the requirements for circuit breaker amperage supplied to bathroom branch circuits. **It is now required that the circuit breaker be the 20-amp type, the 15-amp type is no longer allowed.**

There are two ways to wire bathroom branch circuits:

1. A dedicated **20-amp GFCI** circuit power supply **may power all receptacles and lighting fixtures in a single bathroom**. These devices **may not** be located outside this single bathroom enclosure.

2. A dedicated **20-amp GFCI** circuit power supply **may power outlet receptacles in multiple bathrooms if it only supplies power to GFCI outlets in these bathrooms.** It **may not** power lighting or other devices within these bathrooms.

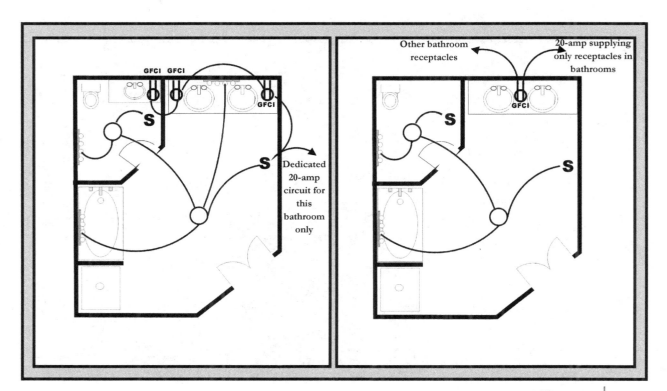

5.5 Planning Circuit Locations

Designing circuit layouts is an important part of installing electrical systems. Proper placement of electrical outlets and even distribution of electrical power allows for safe and functional electrical circuits throughout the structure. **Before you begin, you should determine the number of lighting switches and appliances required to be run throughout the structure.** You will also need to consider if you want audio, video, or communication/internet wiring.

The *NEC®* provides some basic guidelines for planning circuits. The first step is determining where the receptacles and switches need to placed, and the second step is to determine the number and size of the circuits needed for the layout.

You should plan all the circuit types outlined in this chapter. You should determine where the following types of circuits and panels are required before you begin the installation:

- Service Entrance Panels and Subpanels

- Small Appliance Branch Circuits

- General Purpose Branch Circuits

- Individual / Dedicated Branch Circuits

- Bathroom Branch Circuits

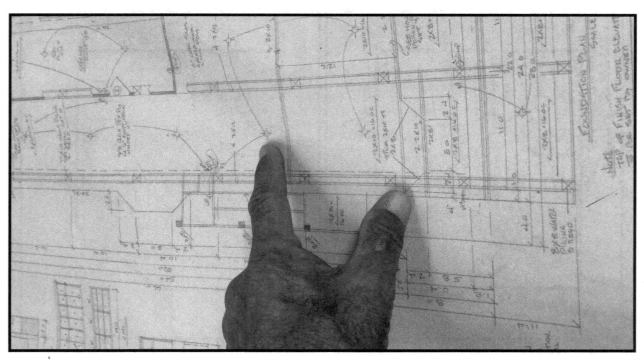

Breaker Panel Working Area

Choosing the location of the breaker panel is the first step in planning circuit diagrams. The locations where the breaker panel can be placed are highly restricted by *NEC®* guidelines to ensure safe operation and that proper spacing around the breaker panel is present.

The **NEC®** **outlines the requirements for panel placement in article 110.26** of the NFPA® handbook. The "working space" around a panel has specific requirements. The **NEC®** **states that the working space must be 6' 6" high, 30" wide, and 3' out from the wall where the breaker is present.** This ensures that there will be no obstruction to the opening of the panel.

Nominal Voltage Transferred to Ground	Minimum Required Clear Distance		
	Condition 1	Condition 2	Condition 3
0-150	3 ft.	3 ft.	3 ft.
151-600	3 ft.	3 1/2 ft.	4 ft.

Condition 1 - Exposed live parts on one side of the working space and no live or grounded parts on the other side of the working space or exposed live parts on both sides of the working space that are effectively guarded by insulating materials.

Condition 2 - Exposed live parts on one side of the working space and grounded parts on the other side of the working space. Concrete, brick, or tile walls shall be considered as grounded.

Condition 3 - Exposed live parts on both sides of the working space.

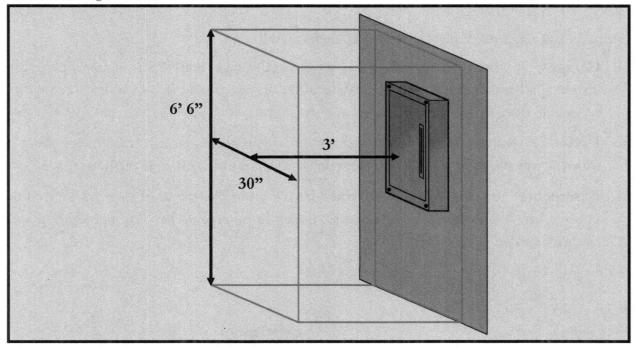

Breaker Panel and Subpanel Locations

Breaker panels should be placed in either residential or commercial buildings where they can be easily reached or accessed. Panels can cause an eyesore in many indoor locations and many people would prefer to have them out of the way. Panel doors or boxes may be painted to match the décor. However, you cannot use anything else to cover the box.

Each panel should be in a well lit place. Not having a well illuminated room where the panel is located can be dangerous. Although it is only required to have a single light in the room where the panel is located, you might consider additional light placements to work safely with the panels.

Breaker panels also cannot have any obstructions such as furniture or major appliances blocking the accessibility of the panel. Breaker panels should also not be placed in enclosed spaces or small areas. These areas include, but are not limited to, cabinets, bathrooms, pantries, closets, or small storage areas.

The easiest way to choose an area for the panel location is to mark the location where enough working space is provided according to the standards put forth by the *NEC®*. **Having enough working space while panels are out of sight and easily accessible will result in a successful circuit planning diagram.**

Panel locations should be central to where they will supply power in the structure. If subpanels are to be used, consider placing it in a location that is central to all the rooms where the subpanel will supply power.

Some of the ideal locations for electrical panels include:

1. **Garages** - If a home or apartment has an attached garage, then this is the ideal place for installing a breaker panel. The placement in the garage should be at eye level and easily accessible from inside the residence.

2. **Under Stairways** - If the proper space is provided, placing an electrical panel under a stairway will allow the panel to be centrally located, out of sight, and easily accessed.

3. **Basements** - Electrical panels will often be placed in basements if they are located in households. It is important that electrical panels in basements have proper lighting and are not exposed to moisture.

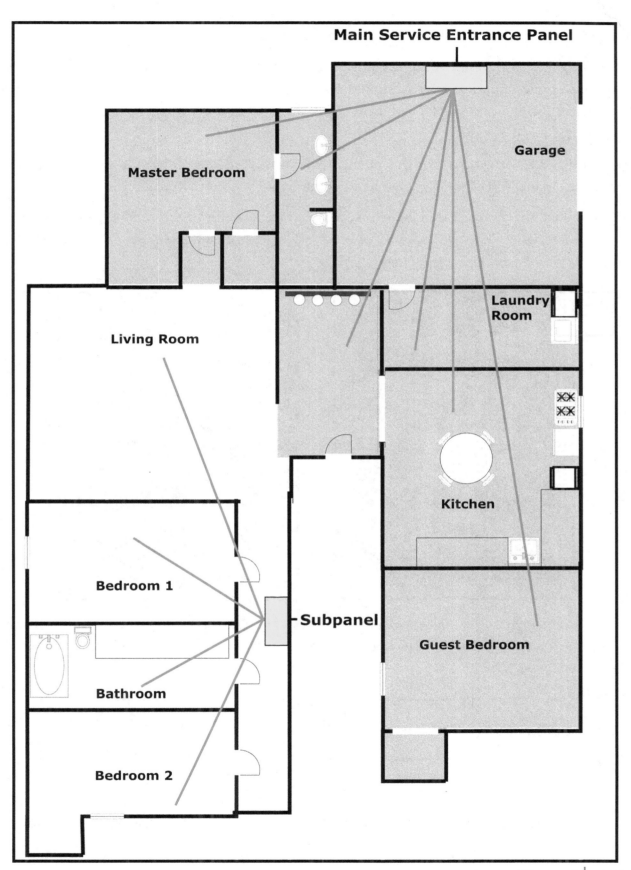

Marking Outlets and Cables for Home Run Circuits

In order to keep track of all the circuit runs in the residential wiring plan, you should mark all of the cables and outlets. You should always designate each circuit with a code number or a letter.

For example, when installing general purpose circuits, we can use the designation GP-1 on the outlets and cable used for this circuit. This procedure is repeated for all other general purpose circuits (GP-2, GP-3, …) and any other branch circuit installations.

You should always properly identify each cable that is run into the service panel with the appropriate label. You should also mark a receptacle or lighting outlet location on the framework for each home run circuit.

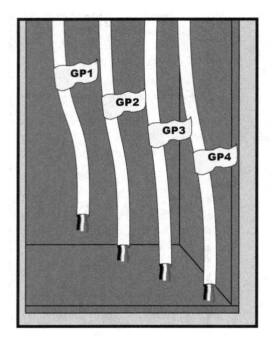 

To determine lighting we use the 3 Volt - Amperes/ft² rule. You must measure rooms from outside to outside including wall thickness. If you have a 22' x 22' room with 2 x 4 walls and 5/8" drywall:

Additional wall length ⟶ 3 1/2" + 5/8" + 5/8" = 4 3/4"

Total Wall Length ⟶ 22' + 4 3/4" = 22.475'

Total Area ⟶ 22.475 x 22.475 = 505 ft²

Total Volt-Amperes ⟶ 505ft² x 3VA = 1515 VA

Therefore one 15A circuit which is rated for 1800 VA (15A x 120V) will be sufficient.

Planning Service Entrance Panel Layout

Proper planning of the layout of circuits connected to the panel is very useful for the installation of circuits in the residence. Having a properly labeled and planned service entrance panel allows for electrical workers and homeowners to easily understand which circuits control certain fixtures, receptacles, and devices throughout the structure.

Planning also helps determine the size of the panel needed. The size is based on the "stab" positions, or places where cables must be inserted into the breakers. **The *NEC®* allows for residential panels to contain up to 60 individual breakers.**

One of the most important benefits is that it allows for load distribution on breakers. This prevents any overload situations that may arise from too many devices being placed on a single breaker.

The panel design below is based on the housing circuit plan of the figure on the following page. Each circuit is either single or double-pole and should be properly labeled on the breaker schedule located on the service entrance panel door or adjacent to the panel.

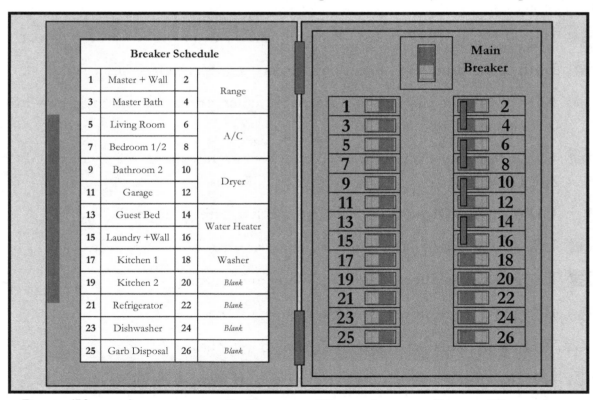

	Breaker Schedule		
1	Master + Wall	2	Range
3	Master Bath	4	
5	Living Room	6	A/C
7	Bedroom 1/2	8	
9	Bathroom 2	10	Dryer
11	Garage	12	
13	Guest Bed	14	Water Heater
15	Laundry +Wall	16	
17	Kitchen 1	18	Washer
19	Kitchen 2	20	*Blank*
21	Refrigerator	22	*Blank*
23	Dishwasher	24	*Blank*
25	Garb Disposal	26	*Blank*

- Range (50 amps)
- Dryer (30 amps)
- A/C (20 amps)
- Water Heater (30 amps)
- Refrigerator (20 amps)
- Washer (20 amps)
- Dishwasher (20 amps)
- Disposal (20 amps)
- Garage (20 amps)
- Kitchen 1/2 (20 amps)
- Bedrooms (20 amps)
- Bathrooms (20 amps)

As seen in this figure, each home run circuit is identified by using letters and numbers to describe where the circuits are installed. <u>These letter designations refer to the panel design we used on the previous page:</u>

GP-1 (Master Bedroom / Living Room Wall) - General purpose circuits that must be AFCI protected.

GP-2 (Living Room / Entrance) - General purpose circuits that must be AFCI protected.

GP-3 (Bedroom 1 and 2) - General purpose circuits that must be AFCI protected.

GP-4 (Garage) - General purpose circuits that must be GFCI protected.

GP-5 (Guest Bedroom) - General purpose circuits that must be AFCI protected.

GP-6 (Laundry Room / Entrance) - General purpose circuits that must be GFCI protected.

Bath-1 (Master Bathroom) - Bathroom circuits that must be GFCI protected.

Bath-2 (Bathroom 2) - Bathroom circuits that must be GFCI protected.

KC-1 (Kitchen Circuits 1) - Kitchen counter receptacles that must be GFCI / AFCI protected.

KC-2 (Kitchen Circuits 2) - Kitchen counter receptacles that must be GFCI / AFCI protected.

GD (Garbage Disposal) - Dedicated circuit to garbage disposal

DW (Dishwasher) - Dedicated circuit to dishwasher (GFCI protected)

Fridge (Refrigerator) - Dedicated circuit to refrigerator

Range (Oven + Stove) - Dedicated circuit to range

WH (Water Heater) - Dedicated circuit to water heater

CW (Clothes Washer) - Dedicated circuit to clothes washer (GFCI protected)

Dryer (Clothes Dryer) - Dedicated circuit to clothes dryer

A/C (Air Conditioning) - Dedicated circuit to air conditioner

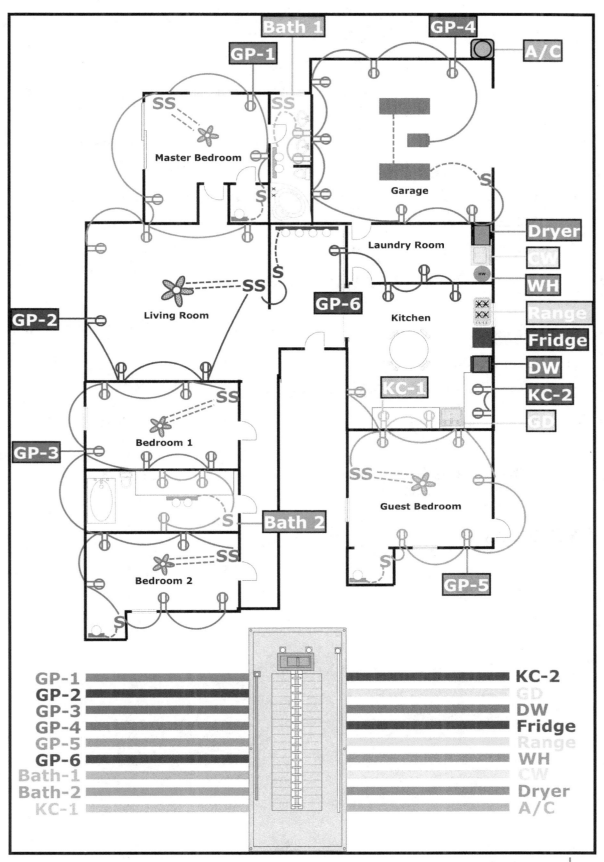

5.6 Chapter 5 Review

1. A path which allows electrons to flow from a power source is called
   _____.

2. The _____ conductor is considered the ungrounded conductor and
   the white conductor is the _____ conductor.

3. A power supply from the main entrance panel to a subpanel is called a
   _____ circuit.

4. The most common type of branch circuit is called _____.

5. Any unbroken wall space of _____ feet or more requires a
   receptacle.

6. Receptacles must be placed on an unbroken wall every _____ feet.

7. Appliances which need their own circuit are called _____ circuits.

8. Circuits used on kitchen countertops are called _____ circuits.

9. A dedicated circuit of _____ amps must be installed for each
   bathroom.

10. Range/Dryer power feeds must use _____ conductor cable.

CHAPTER 5 NOTES:

Electrical Service

The electrical service is the complete wiring system from the utility company's overhead or underground distribution lines to residential or commercial service panels in individual structures.

The "service conductor" refers to the conductors that are provided through the utility company's distribution lines run to a building. The service entrance conductors are installed by electrical contractors.

In this chapter we will discuss the steps to installing service conductors for use through the service panels. This is the first step in bringing electrical power to a structure and also deals with the largest electrical current since these conductors are used to supply power to an entire structure. Make sure to follow correct safety guidelines to prevent accidents.

The various procedures for installing electrical service are:

1. Service drop (overhead)

2. Installing service entrance conductors and equipment

3. Service laterals (underground)

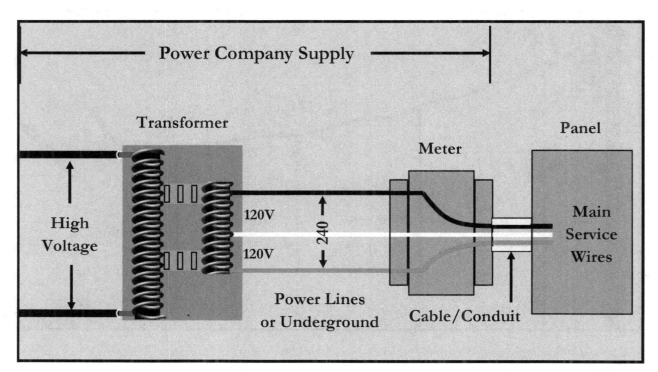

6.1 Service Drop

In the United States we use the 120/240V split-phase system for the overhead drop in residential installations. Several residences in the same area are usually powered through the use of a transformer distribution system which is often mounted on an electrical pole.

Each service drop consists of two 120V lines and a neutral line. Together they are referred to as triplex cable. If you are running the triplex cable over large distances then supporting messenger cables may be used to help stabilize the line.

The neutral line is run from the electrical pole to the neutral bar in the service panel. This neutral bar is connected to a grounding device or conductor. This grounding conductor is part of the grounding system and is usually attached to a grounding rod which is inserted into the ground itself. There are two 120V lines (phase A + phase B) which measures 240V between the lines. If only 120V is needed then you can use a single 120V phase line.

Commercial service drops are usually much bigger and contain a three-phase system. These systems are usually 120Y/208 (three 120V circuits 120 degrees out of phase with a 208V line to line), a 240V three-phase, or a 480V three-phase.

The *NEC®* requires that service drop conductors with ratings of 600 volts or less be installed with certain clearance requirements above the ground grade.

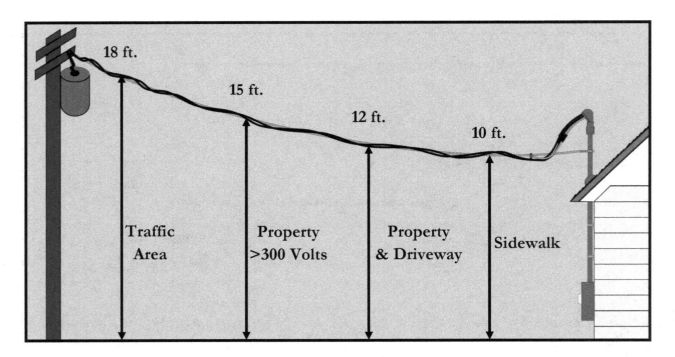

It is important that steps are taken to prevent damaging of the service conductors. Since the service conductors are installed outdoors it is important that you protect them from moisture. The area that is most often affected by moisture damage and rain is the raceway service entrance.

The point of attachment for the service conductors should be at least 6 inches below the termination point. This means that the connections from the service entrance conductors to the service drop conductors will follow a downward path.

If you cannot place the point of attachment below the termination point, then a drip loop must be used. This allows the point of attachment to be above the raceway entrance. The drip loop is formed where the service entrance cables are allowed to hang below the raceway entrance and the point of attachment is connected to a wall with an insulated ring or locking device.

The bottom left picture shows the point of attachment 6 inches below the termination point. The picture on the right shows that the attachment point is above the termination point, but a drip loop is created to protect from moisture in the raceway.

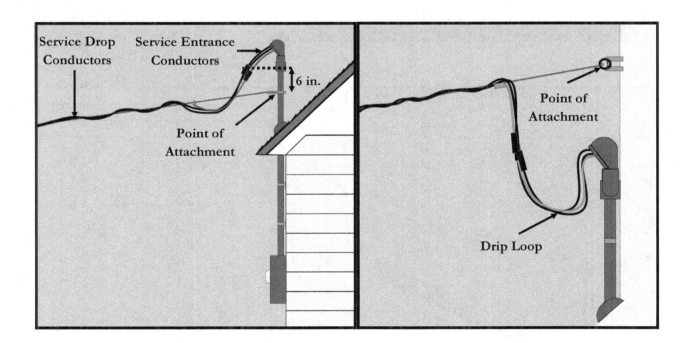

Service conductors run overhead must have a **minimum vertical clearance of 8 feet above the surface of a roof if the slope is less than 4 inches for every 12 inches of roof. A vertical clearance of 8 feet is also required for service entrance conductors that are 300V or more**. If the roof is sloped more than 4 inches for every 12 inches of roof, then the 120/280V or 120/240V conductor clearance can be lowered to 3 feet over the roof.

The attachment point for the service conductors cannot be closer than 3 feet to windows, porches, fire escapes, or any place easily accessible to people.

The service conductor entrance should not be located more than 4 feet above the edge of the roof for safety reasons. In addition, the drip loop should be located a minimum of 18 inches above the surface of the roof. This allows enough space for servicing the conductors without potentially damaging the lowest hanging portion of the conductors.

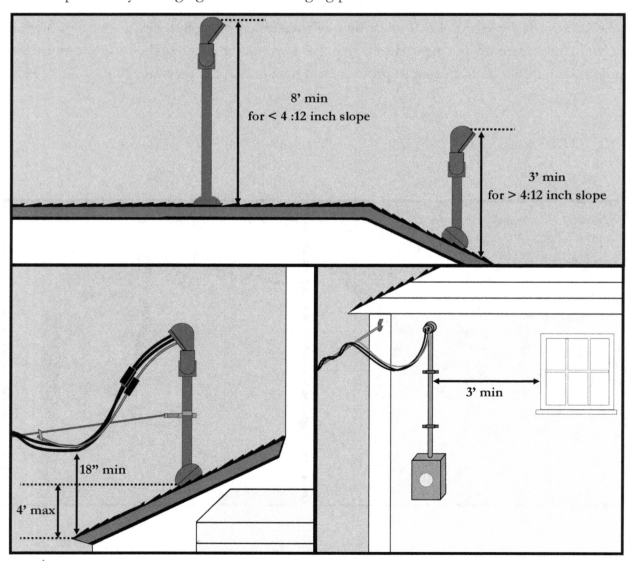

6.2 Service Entrance Equipment

The service entrance is an assembly of parts designed to easily supply service entrance cables to the residence. These parts can include the entrance head, which may come in a variety of types or styles, the service mast which is either conduit or SE cable without conduit, the meter box, and the connections used to attach the parts together.

The entrance head, also known as the weather head, is the device used at the top of the service entrance mast. They are used with either conduit installations or SE cable installations. There are two types of entrance heads depending on whether the service conduit has threads or not.

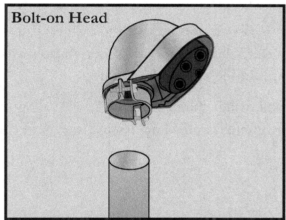

The **bolt-on clamp entrance head** is used for threadless conduit and is attached by tightening the screws on the clamp. Make sure that the head is secure after tightening the screws on the clamps.

Threaded service entrance heads are used with threaded conduit and are attached by twisting the entrance head onto the conduit.

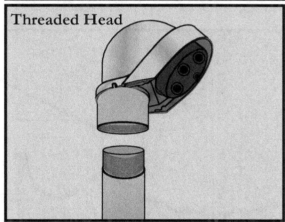

The **conduit** used for the service installation can be RMC, IMC, EMT, or EPVC. Each conduit type provides a frictionless inside surface to easily feed the service cables through.

The **conduit connectors** attach conduit to the surface of the meter box. These connectors must be UL approved and may come in various sizes depending on the type of conduit used. The connectors are secured with screws attaching the conduit to the holes in the meter box.

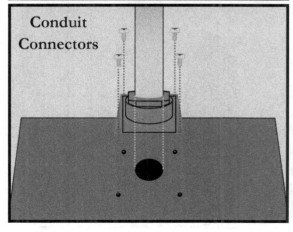

6.3 Installing Service Entrance Equipment

Meter installation locations will usually be determined by the utility company and should be placed in easily accessible areas and not behind any type of fence or barricade. It is recommended that the **meters be placed at least 4 feet from the ground** to protect from environmental damage, but **not above 6 feet** to allow any service person easy access from the ground.

It is also important that the meter location be free from obstructions in all areas around the meter. This includes a **3 feet radius around the meter where no foliage or outdoor appliance should be located.**

Following these guidelines will allow utility company employees to provide proper maintenance, reading, testing, and inspection of the meter when needed.

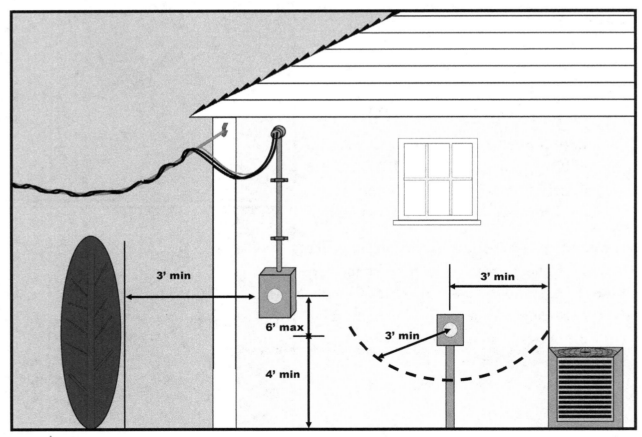

6.4 Installing Service Entrance Conductors

Before you begin, make sure to measure the length of conductors needed for the installation. After measuring the correct length, cut the conductors with an added 4 feet of conductor for any connections that are needed at the entrance head and the meter. You should allow an extra 3 feet at the top of the mast and an extra foot of conductor in the meter box.

Start by inserting the service entrance conductors into the top of the conduit with no entrance head attached.

Continue to feed the conductors through the top of the mast until they reach the meter connection.

You should continue to feed the conductors until you have the extra foot of service entrance conductors to work with in the meter box.

Once the required amount of service entrance conductors have been threaded through the conduit you should attach the service entrance head.

Make sure the mast head is securely attached and that you have 3 feet of conductor to use for the connection to the service drop conductors.

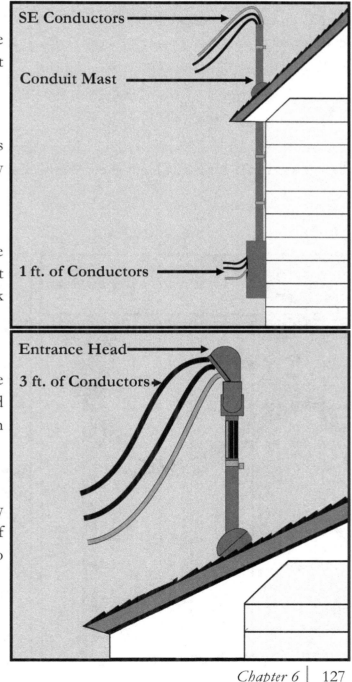

6.5 SE Cable Installations

Using SE cable for service installation allows you to install the service without the use of conduit. In order to make the connections from the SE cable to the service drop connectors, **you must remove at least 3 feet of the service entrance conductors from the SE cable** by cutting the outer jacket.

Once you have measured the proper length of SE cable needed to make the meter connection, cut an additional 3 feet of cable to expose the SE conductors. Feed these conductors through the service head until you have the proper length of conductor necessary for service connections.

Re-attach the entrance head to the top of the SE cable by tightening the clamp and securing it to the surface of the wall.

Secure the SE cable in the meter box with waterproof fittings. **The SE cable should also be secured with straps at least every 20 inches.**

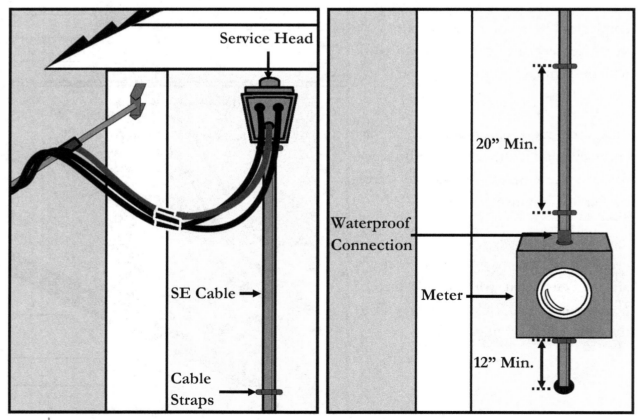

6.6 Service Laterals

Many service installations today are done with underground service laterals. These underground service laterals can be used for residential and commercial applications. Having the service entrance conductors run underground prevents possible damage from the elements.

These service laterals can be run from either a pole-mounted transformer, or from a pad-mounted transformer. The cables to and from the pad-mounted transformer are underground.

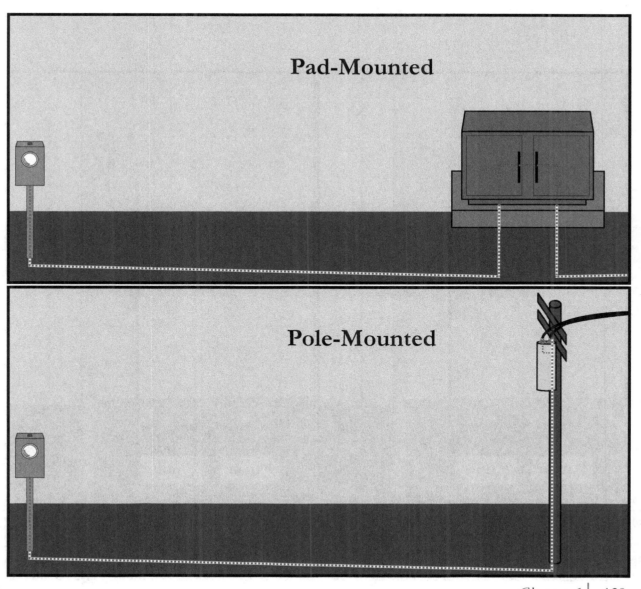

Although these underground laterals are protected from the elements by being underground, they can still be damaged by moisture if they are not properly installed. The cables or conductors used for these laterals must be designated as Underground Service Entrance (USE) materials. Conduit is also used if there is a higher chance of the lateral being damaged.

There are two methods for connecting the service lateral to the meter base. Both methods require that the service lateral be installed at least 2 feet below ground and no more than 4 feet. Some local codes may require the use of a 45° angle to allow the conductors to enter the conduit below the grade.

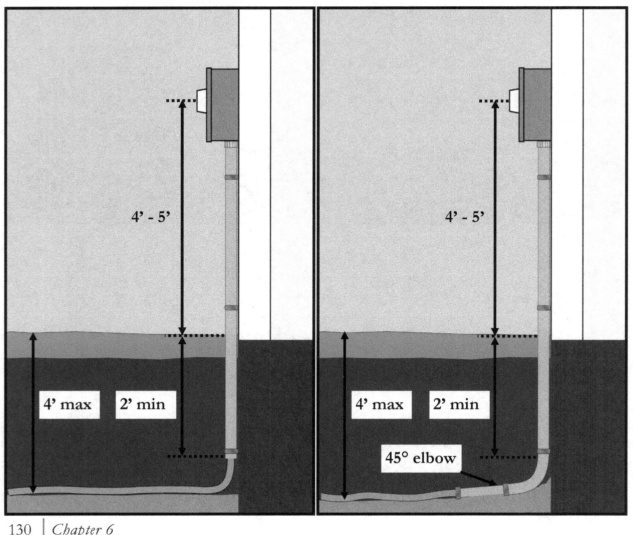

6.7 Meter Base Connections

There are two types of meter connections. On the left is the meter connection used for underground lateral service entrance conductors. The right diagram depicts the connections made for overhead service entrance conductors.

It should be noted that in the overhead service connection diagram that both cables may have two black conductors and one white conductor.

The grounding electrode conductor should be connected to the base of the meter box plate. **It is important to ensure that this conductor does not touch the load conductors. This can cause a malfunction of the box and will cause an accident.**

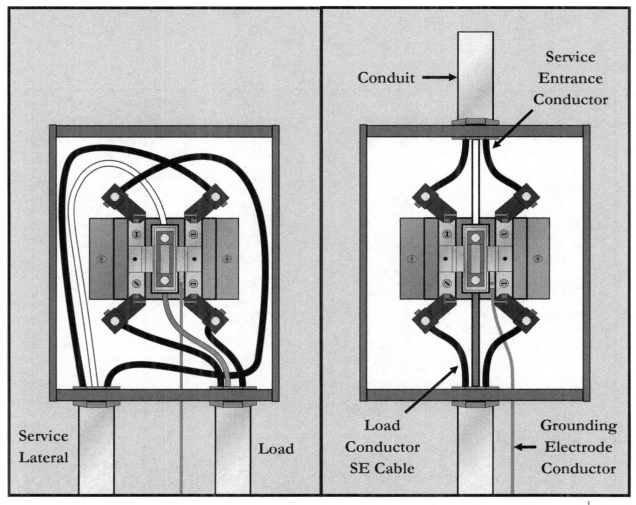

6.8 Meter to Panel Connection

Routing SE cable through the wall

After connecting the SE cable to the meter box, **make a hole at least 12 inches below the meter box into the structure where you will be running the cable.** Run the cable from the bottom knockout of the meter into the hole.

Run the SE cable from the hole to the electrical panel box through the structure. This is usually done by running it through the floor joist or basement area. Ensure that the SE cable is not placed in an area where it can be damaged.

The SE cable should be secured to the framework with straps. The SE cable should be run through the bottom of the service entrance panel and be secured with a SE cable clamp to the electrical panel.

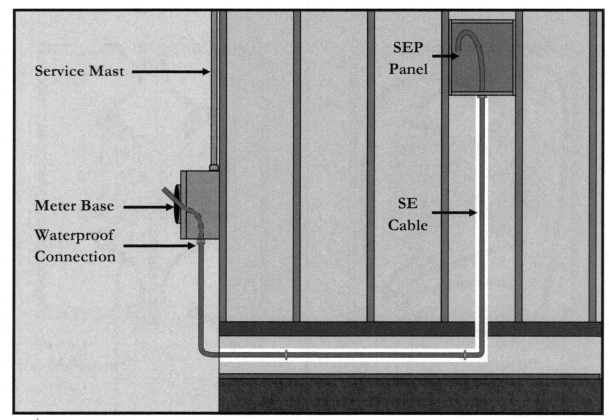

Routing SE cable through conduit into panel

After the conduit from the service entrance to the meter box has been installed you should use the bottom knockout on the meter box to attach a conduit body to run conduit to the electrical panel box.

Make the correct waterproof connections from the meter base to the SEP panel. Run the SE conductors from the bottom of the meter base to the SEP panel through the conduit.

Once the conduits have been run into the SEP panel, make sure to install the proper grounding bushing with bonding jumper to the SEP knockout where the conduit is connected.

Connect the bonding jumper from the bushing to the grounding bus in the panel.

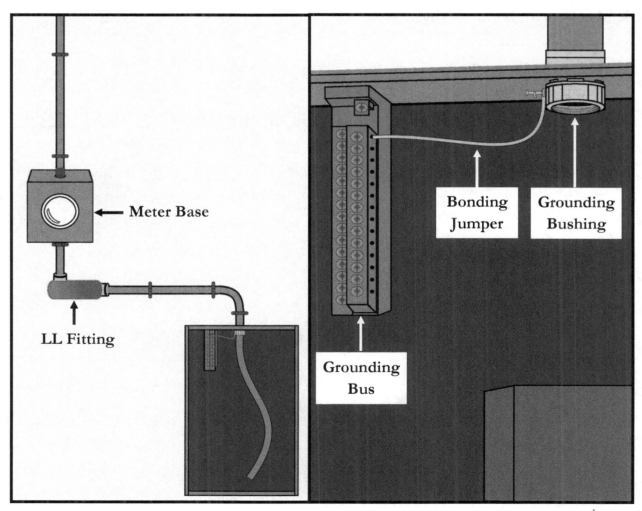

Meter Base

LL Fitting

Bonding Jumper

Grounding Bushing

Grounding Bus

6.9 Chapter 6 Review

1. The service entrance conductors are installed by the _____.

2. Overhead service entrance conductors run from a pole-mounted transformer are required to be at least _____ feet above the sidewalk.

3. The service entrance conductors drip loop needs to be at least _____ inches above the roof.

4. The electrical meter must be between _____ and _____ feet from the ground.

5. You must leave at least _____ feet of service conductors from the service entrance head for utility hook up.

6. SE cable should be strapped every _____ inches.

7. The two ways transformers are mounted are _____ or _____.

8. The wires from the meter to the SEP are connected to the _____ terminals of the meter.

9. Service lateral conductors must be buried from _____ to_____ feet below ground.

10. SE cables must be secured no less than _____ inches from the meter.

CHAPTER 6 NOTES:

Panels and Breakers

7

Breaker panels are also known as service panels, electrical panels, load centers, or breaker boxes. These panels are a sheet metal or steel box which house circuit breakers. These **circuit breakers are wired to circuits which are distributed throughout a structure. They also provide overcurrent protection through their breakers/fuses.**

There are various service entrance panels and subpanels used in residential or commercial wiring installations. Depending on their intended use, panels come in a variety of styles, sizes, and designs. Remember that when dealing with panels that they must meet both *NEC®* and local code regulations.

In this chapter we will discuss topics surrounding panels and breakers including:

1. Service Entrance Panel Parts

2. Service Entrance Panel Types

3. Installing SEP

4. Subpanels

5. Breaker Types

6. Calculating Panel Ratings

7.1 Service Entrance Panel Parts

Service entrance panels provide a way of connecting the service utility to the residential wiring system. They also offer overcurrent protection for branch or feeder circuits.

The main service entrance panel parts include:

1. **Main Lug Terminal** - The main connection terminals for the ungrounded utility service entrance conductors.

2. **Neutral Conductor Terminal** - The neutral bar used to connect the service entrance grounded conductor.

3. **Bonding Strap** - Used to bond the grounding bus bar to the service entrance panel.

4. **Neutral Bus Bar** - Used to terminate all of the equipment grounding circuit's conductors in the service entrance panel.

5. **Grounding Bus Bar** - Used to terminate all of the equipment grounding conductors in the service entrance panel.

6. **Bonding Jumper Bar** - Bar used to connect the neutral bus bar and the grounding bus bar to the grounding electrode system.

7. **Grounding Electrode Terminal** - The bottom terminal located on the neutral bus bar that is used to connect the wiring system to the grounding electrode conductor.

8. **Main Breaker Switch** - The main breaker switch is the main disconnect to control electrical flow from the service entrance conductors to the other breakers on the panel and the wiring system. In typical installations there are two main breaker switches for each side of the power bus bar.

9. **Single-Pole Breaker** - Circuit breaker overcurrent protection device to protect a circuit and control a current for a 120V branch circuit.

10. **Double-Pole Breaker** - Circuit breaker overcurrent protection device to protect a circuit and control a current for a 240V branch circuit.

11. **Power Bus** - Equipment used to conduct the electrical current through the main breakers supplied by the service entrance conductors. Each side of the power bus is controlled by a main breaker switch. The available positions are called stabs.

12. **Grounding Electrode Conductor** - Connects the electrical wiring system to the grounding electrode system and bus bar.

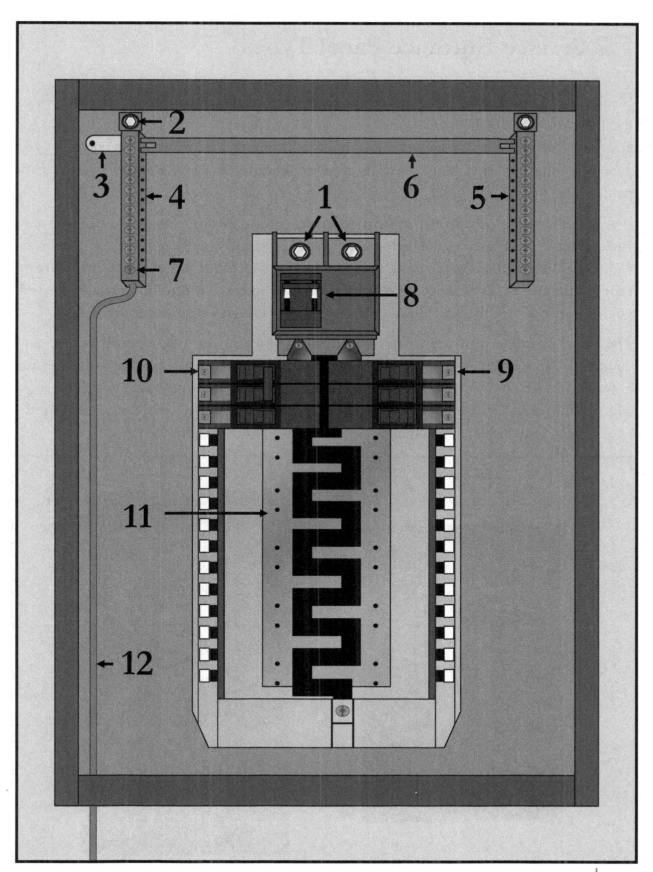

7.2 Service Entrance Panel Types

Electrical panels are usually set up with 2 circuit breaker columns that are operable from the front. Some panelboards come with a door covering the breaker switch handles for greater protection, but all panels come with a dead front. A dead front (cover) prevents any person who is working on the panel from coming into direct contact with live electrical parts.

Bus bars are used to carry the current from the hot conductors to the breakers. These current lines are secured to bus bars with either a threaded screw bolt-on connection or with a retaining clip, also known as a plug-in connection. **Most panelboards use screw-type connections, especially in commercial and industrial applications. However, you may encounter plug-in connections when working on residential structures**.

The most common size panels in residential applications are 100-amp, 150-amp, and 200-amp. Using a 200-amp panel will provide enough electrical power and protection for homes up to 3,500 sq. ft. Using a 200-amp panel will ensure that you have enough room to add switches and circuit breakers to the residence. For large homes exceeding 3,500 sq. ft., two 200-amp panels may be needed or an additional 100 or 150-amp panel.

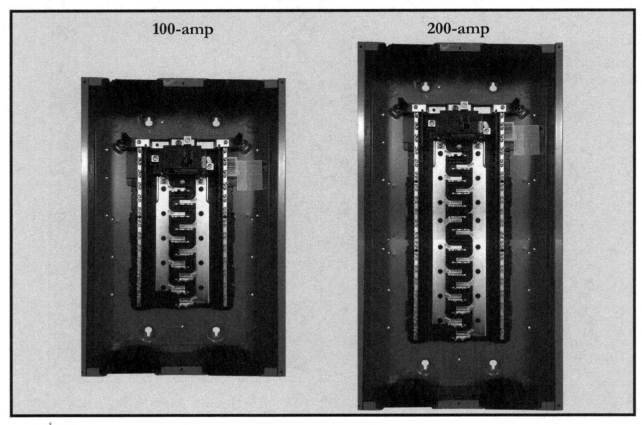

100-amp   200-amp

Service entrance panels can be either main breaker or main lug only. **Main breaker panels have a single main breaker that can be used to cut all the power to the breakers** and household. The main breaker is a double-pole circuit breaker that not only protects the other circuits connected to the panel, but also controls the flow of energy from the service entrance conductors. Main breakers are permitted to be installed when both the meter and feeder cable are within 10 feet of the panel itself. Some local codes may require that the panel be even closer.

Main lug panels do not contain a main breaker. The conductors are instead run to the main lugs. Because of the direct connection to the lugs, a separate disconnect is required. This disconnect is typically installed in the form of a main breaker at the location of the meter. For subpanels using lug type connections, the separate disconnect can be considered the main breaker at the service entrance panel. This type of installation is beneficial during major emergencies where authorities or emergency response personnel need to disconnect the power from the outside.

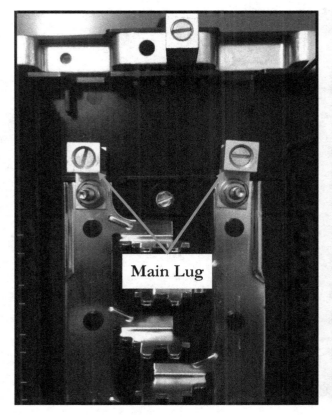

Damp and Wet Locations

NEC® guidelines in article 408 cover the requirements and codes related to switchboards and panelboards. When the *NEC®* refers to the panelboard, they are referring to the inside of the cabinet where the electrical parts of the panel are located. The outside enclosure that surrounds the panelboard is called the cabinet. **Guidelines and regulations for cabinets can be found in *NEC®* article 310.**

All cabinets for panelboards must effectively prevent moisture and water from entering the cabinet. **If cabinets are installed in wet locations it must be designated as waterproof as per *NEC®* guidelines outlined in 312.2.**

If the cabinet is mounted on a surface in a wet location then it must be mounted with at least 1/4 in. of space between the cabinet and the mounting surface to prevent any moisture from the surface reaching the inside of the enclosure. **Regulations for surface mounting in wet locations can be found in *NEC®* codes 312.2 and 408.37.**

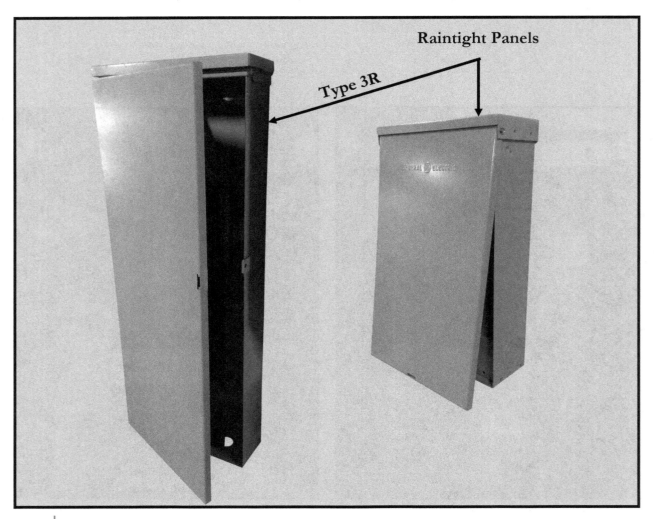

Enclosure Types

Enclosures (cabinets) are given letter and number designations referring to their ability to prevent environmental damage and intended installation locations. They may be designated as type 1, 2, 3, 3R, 3S, 4, 4X, 5, 6, 6P, 12, 12K, or 13. They can also have word designations such as "raintight" or "rainproof".

Type	Location	Installation and Intended Use
1	Indoor	Provides limited protection against dirt and dust.
2	Indoor	Provides limited protection against dirt, dust, and water.
3	Outdoor	Provides protection against windblown dust, rain, sleet, and external ice formation.
3R	Outdoor	Provides protection against rain, sleet, and external ice formation.
3S	Outdoor	Provides protection against windblown dust, rain, and sleet and allows for continued external operation even surrounded by ice.
4	Indoor / Outdoor	Provides protection against splashing water, windblown dust and rain, hose-directed water, and external ice formation.
4X	Indoor / Outdoor	Provides protection against splashing water, corrosion, windblown dust and rain, hose-directed water, and external ice formation.
5	Indoor	Provides protection against settling dust, falling dirt, and dripping noncorrosive liquids.
6	Indoor / Outdoor	Provides protection against hose-directed water, invasion of water during temporary submersion, and external ice formation.
6P	Indoor / Outdoor	Provides protection against hose-directed water, invasion of water during prolonged submersion at a certain depth, and external ice formation.
12, 12K	Indoor	Provides protection against circulating dust, falling dirt, and dripping noncorrosive liquids.
13	Indoor	Provides protection against dust and spraying water, oil, and noncorrosive liquids.

Barrier Protections

Recent changes to codes surrounding installation of panelboards and switchboards now require the use of barriers. These barriers are placed within the enclosures which creates a layer of protection from exposure to ungrounded bus bars and service terminals. The *NEC®* has mandated that the use of these barriers be put into practice due to safety concerns when maintenance is being conducted on live parts of the electrical panel.

These barriers create an insulation layer on the service side of the uninsulated parts of the panel. The protection provided by these barriers is easily achieved when panelboards are designed for a single-service disconnect. Installing these barriers allows for codes established in the NFPA® 70E to be met as necessary safety parameters for performing electrical work on panels.

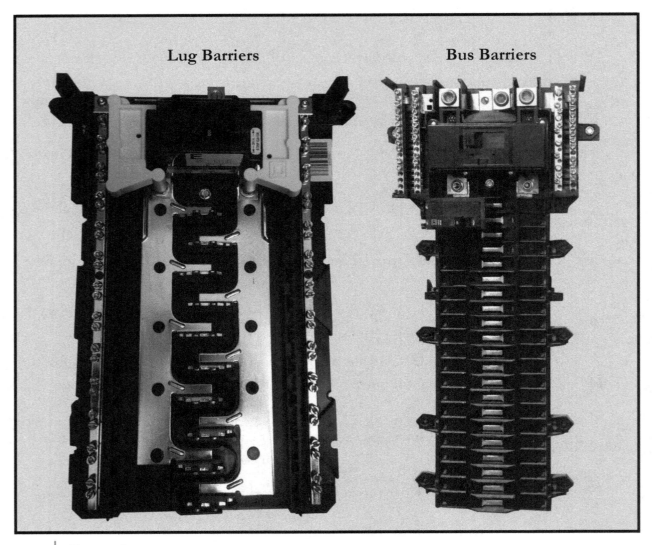

Lug Barriers   **Bus Barriers**

Panelboard Regulations

According to standards put forth by UL® Standard 67, all panelboards used for lighting and appliances must be class CTL (Circuit Total Limiting). This regulation is in line with the *NEC®* regulation which requires that each panelboard must be designed to physically prevent installation of excessive overcurrent devices for which the panelboard was not designed or rated to handle.

When we refer to a panelboard with a 20-40 load center, we are talking about the 20 spaces for full size breakers or available space for tandem breakers for a total of 40 breakers. CTL panelboards use a "rejection feature" to prevent additional circuit breakers from being installed. This rejection feature prevents tandem circuit breakers from being installed in locations which are prohibited by *NEC®* and UL® standards.

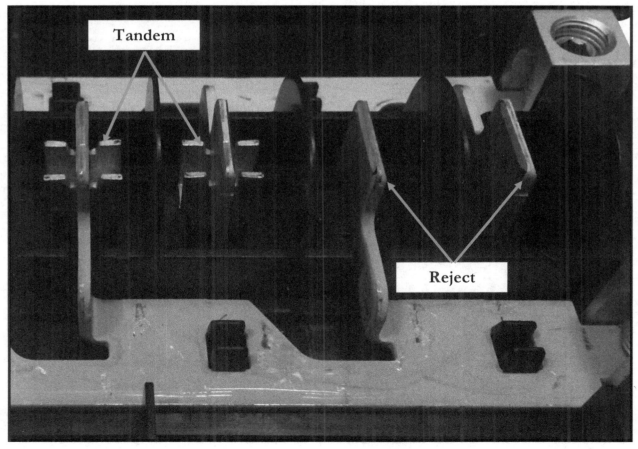

7.3 Installing Service Entrance Panels

The first step to installing electrical systems in residential or commercial structures is to connect the service entrance panel to the service entrance conductors.

<u>The steps to connect the SEP panel are:</u>

1. If the bus bar jumper did not come pre-installed, then connect it between the grounding bus bar and the neutral bus bar.

2. Connect both hot service entrance conductors to the main breaker terminals. The hot conductors can be connected using either an Allen wrench or in some cases screwdrivers. Ensure that the service entrance conductors are securely connected.

3. Connect the neutral service entrance conductor to the lug on the neutral bus bar.

4. Bond the grounding strap to the service entrance panel. If the bus bars are connected with a jumper, then the bonding strap may be connected to either bar along with a green screw. If they are not connected, use the grounding bus bar for the bond.

5. Connect the grounding electrode jumper to the neutral bus bar.

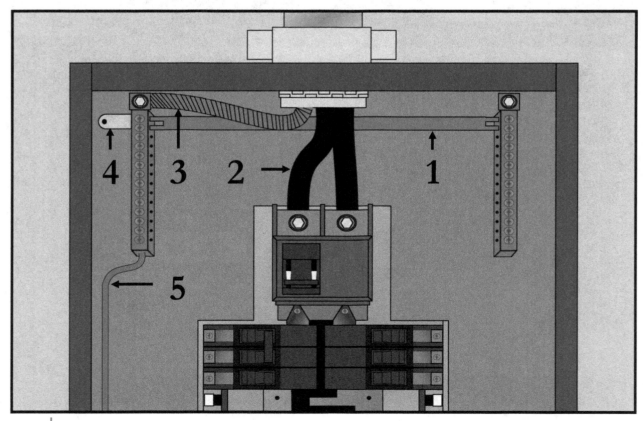

It is important that when you are connecting devices to your electrical panel that you plan the paths for the conductors. **The heat that is created over time must have enough space to escape the panel.** In addition, the wires that are meant to ground devices cannot touch hot conductors or the hot power bus.

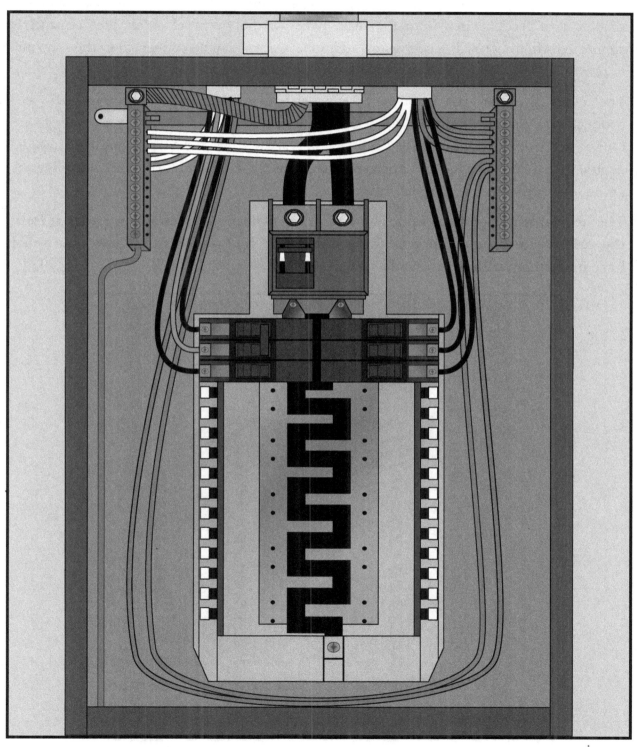

7.4 Subpanels

Subpanels are used to supply power and overcurrent protection in areas that are far from the main service entrance panel. This can also simplify circuit wiring plans by running a single set of feeder wires from the main panel to the subpanel. After the subpanel is connected with these SEP feeder wires, additional circuits can be connected to the subpanel to power outlets, lighting, and other individual circuit breakers.

The feeder wires that run from the SEP panel to the subpanel are comprised of 2 hot conductors, a neutral conductor, and a grounding wire. The neutral and the grounding wire cannot be bonded to the subpanel. This is only allowed for the SEP panel. Bonding the neutral wire to the subpanel can open multiple paths for the current to return, including the grounding wires which is extremely dangerous.

The neutral bar also needs to be isolated using a plastic insulator to separate it from the subpanel cabinet. The grounding bar should be bonded to the case and must have its own grounding electrode system.

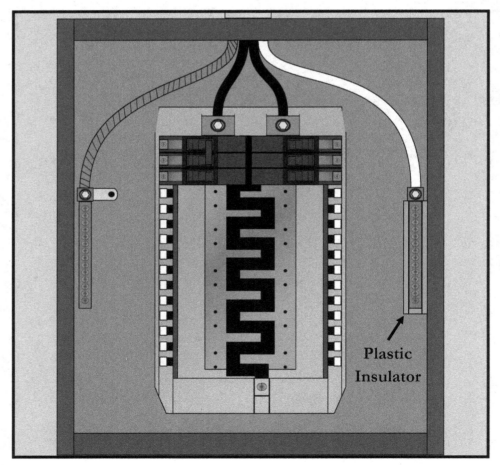

Plastic
Insulator

7.5 Circuit Breakers

A circuit breaker is a mechanical device used to make and break a connection from the feeder circuit to individual branch circuits. These devices protect wiring from damage by "tripping" when an electrical short or overcurrent occurs. The tripping mechanism physically breaks the connection between the power supply and the circuit connected to the breaker.

When this trip occurs, the lever on the front of the breaker will snap to a position between the OFF and ON position. In the case of AFCI or GFCI breakers, there will be a reset button. Some other circuit breakers may have a small peep sight that is red when the circuit breaker has been tripped. These mechanics allows for clear indication of which breakers are experiencing problems.

Thermal Magnetic Circuit Breakers

Thermal magnetic circuit breakers are the most common type found in electrical panelboards. They use two techniques simultaneously to protect circuits. The first is a metallic strip to protect against lower overcurrent surges that may occur over long periods of time. The second is an electromagnet which protects against sudden large surges, also know as shorts. The thermal feature of the breaker allows for a timed trip in which large surges are cut quickly, while smaller and longer overcurrents take longer to trip. This prevents small temporary spikes of overcurrent from tripping the breaker when motors or large applications are turned on. However, large surges will trip the breaker immediately.

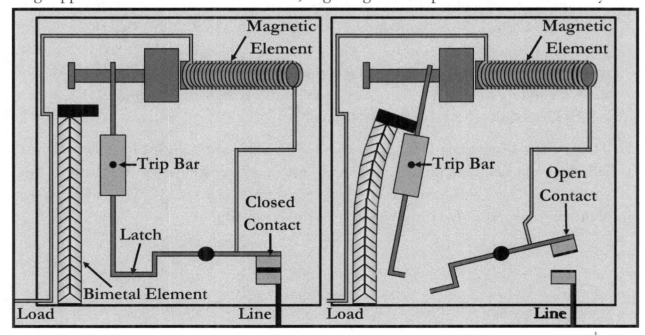

Circuit Breaker Amperage Rating

Each circuit breaker is rated to a specific amperage. When this amperage is exceeded the circuit breaker will trip. If you take the time to calculate the rated amperage and the anticipated load, you can prevent the possibility of overcurrent tripping.

Each individual breaker should have a number mark for the maximum allowed amperage. This is the maximum allowed amperage before the circuit breaker will trip.

As a rule of thumb, you should not exceed 80% amperage load for each circuit breaker. This provides enough amperage for short surges from appliances and devices.

For typical 20A circuit breakers:

Amperage x 0.8 ⟶ 20 amps x 0.8 = 16A allowable load

Circuit Breaker Types

1. **Single-Pole Breakers** - SP breakers are used for low amperage household use. SP breakers are rated at 15, 20, or 30A at 120V. They are used for small appliances such as light power tools, typical household lighting, and devices plugged into living area receptacles. SP breakers are connected with a single hot. When the breaker detects an overcurrent surge in the hot conductor, the breaker trips, cutting off the hot conductor connection.

2. **Double-Pole Breakers** - DP breakers are used when larger voltage and amperage ratings are needed. Most household DP breakers are rated at 20-60A at 240V. DP breakers are used for water heaters, central air conditioning units, electrical ranges, and electric clothes dryers. DP breakers are connected with two hot conductors. If either of the hot conductors shorts or overloads it will trip the breaker as a whole. DP breakers can also be used to protect two 120V circuits.

3. **Triple-Pole Breakers** - These breakers are used to supply power to commercial or industrial appliances and motors. These breakers are connected using three hot conductors. These breakers have large amperage and voltage ratings. When these ratings are reached on any of the three phases the breaker will trip.

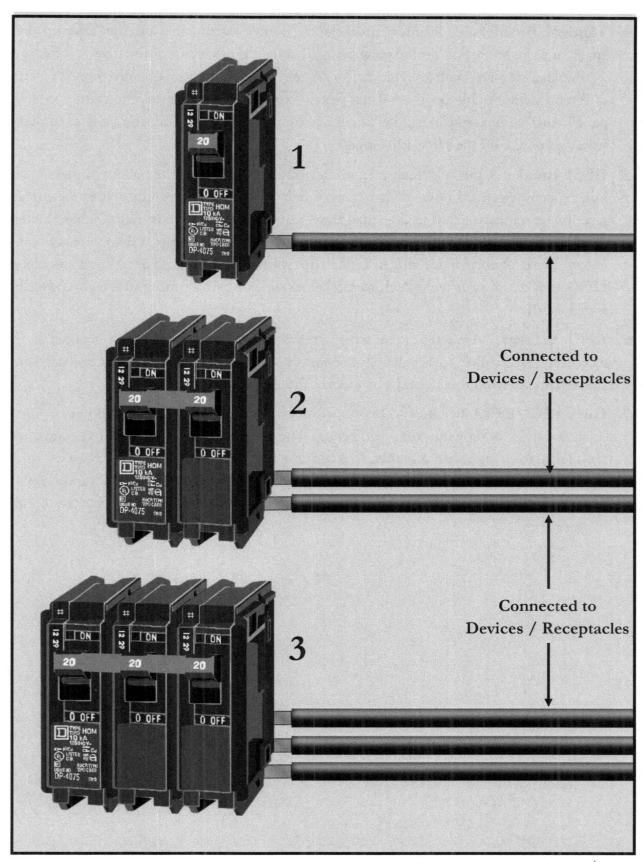

1

2

Connected to
Devices / Receptacles

3

Connected to
Devices / Receptacles

4. **Tandem Breakers -** Tandem breakers are double circuit breakers that take up the space of a single breaker on the panelboard. Tandem breakers are sometimes referred to as duplex, slimline, half-weight, half-inch, or twin breakers. Tandem breakers are different than double-pole breakers which connect to two different poles on the panelboard with a single handle tie. Each of the two hot conductors on a tandem breaker operate on the same 120V phase.

5. **GFCI Breaker -** Ground-fault circuit interrupter breakers have the ability to detect an imbalance in electrical flow when a hot conductor touches a ground, such as the metal layer on an appliance. These abnormal flows are often referred to as faults. These faults can transfer electrical flow through people and cause serious injury. GFCI breakers can detect these abnormalities much faster than normal breakers. The *NEC®* requires GFCI breakers in wet areas such as kitchen counters, bathrooms, and attached garage locations.

6. **AFCI Breaker -** Arc-fault circuit interrupter breakers can quickly detect small arcs of electrical current due to accidental disconnects of hot conductors. They are used to prevent an arc from causing fires by quickly shutting them down.

7. **Dual AFCI/GFCI Breaker -** Dual-function circuit breakers use the function of a class A 5mA GFCI along with AFCI protection. These breakers protect against ground and arc-faults, and come with an additional self-test feature. The *NEC®* now requires that breakers installed for kitchen and laundry applications be both AFCI and GFCI protected. Using dual-function circuit breakers to meet these standards reduces cost and maintenance.

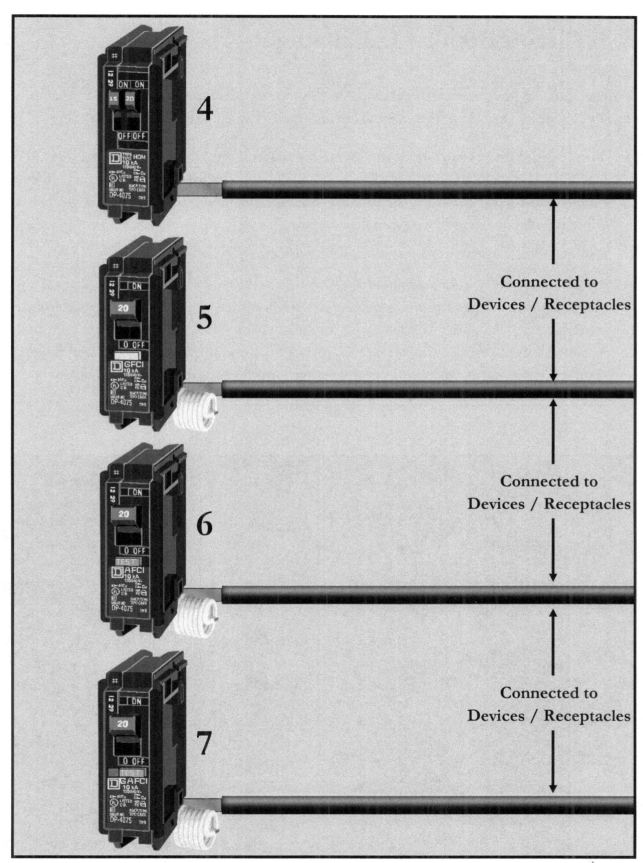

4

5

Connected to
Devices / Receptacles

6

Connected to
Devices / Receptacles

7

Connected to
Devices / Receptacles

7.6 Circuit Breaker Installations

Proper installation of circuit breakers is an important part of setting up your electrical panels. The following are steps to install circuit breakers safely and effectively:

1. **Turn off the power supply to the electrical panel.** Flip the main breaker panel switch to OFF. The main breaker switch can be at the top or bottom of the panel. If you cannot find the main circuit panel switch, then a separate disconnect may be located in another panel or separately housed disconnect.

2. Remove the electrical panel cover from the cabinet.

3. **Make sure power is disconnected.** Use a high rated AC voltage meter to detect if any power is being supplied to the panelboard. Use the tester by touching one probe to the main lug and the other to the screw terminal where there is a black, red, or blue insulated wire connected. If the meter detects 120V or more then the panel is still being supplied with power. If there is power, do not continue until all the power has been properly disconnected.

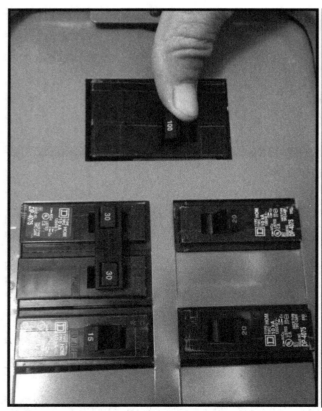

4. Locate an unused location space around the other circuit breakers. **A single open space will provide a single 120V circuit. In the case of a tandem breaker, it will provide two 120V circuits. A double-pole breaker attached to two open spaces will provide a 208V or 240V circuit.** Make sure that the intended space for the new breaker is compatible with the panelboard cover. The cover must have the ability to access the face of the breaker once it is reconnected.

5. Choose a circuit breaker type. Each panel will list the compatible circuit breaker types that can be installed. **If you do not follow the panel label, and install unapproved breakers, this will void any UL approved usage of the electrical panel.** If possible, breakers that are installed should be from the same manufacturer as the electrical panel.

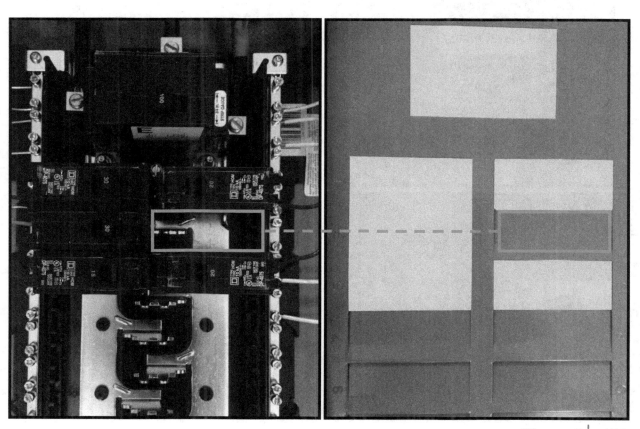

6. Locate the mounting points where the breaker is to be installed. The circuit breaker will have two mechanical contact points. The first contact point is a nonelectrical point at the screw terminal end of the circuit breaker. This contact point can be a clip, tang, or bracket which acts as the support for the outboard part of the panel and should be connected first. The second electrical contact is on the opposite side of the first mounting point and should be installed last. The contact is made inside of the circuit breaker where it is barely visible. This connection is complete when the breaker is flush to the bus bar.

7. Make sure the breaker is in the OFF position. There are three settings for the handle of the breaker. There is the ON, OFF, and tripped position which is in the middle.

8. Install the breaker in the designated space on the panel. Simply apply downward pressure on the breaker to make the connection.

9. Connect the conductors. Ensure that the breaker is still switched off. Connect the conductors to the circuit breaker terminals, ground terminal screw, and neutral bar terminal screw. Install the cover and test the power being supplied to the breaker.

7.7 Service Load Calculations

It is necessary to calculate how much power is needed to supply the electrical appliances connected to each circuit breaker. As the electrical field continues to advance, we will see higher loads and breaker ratings being installed. **Below are theoretical installation calculations for determining the necessary service amperage needed following NEC® code 220.83A.**

Service Load Calculations		
1600 sq. ft. Space	**Ratings**	**Total Load**
Light and Convenience Outlets	3 VA x 1600 sq. ft.	4800 VA
Small Appliance Circuits	3 x 1500 VA	4500 VA
Laundry	1 x 1500 VA	1500 VA
Electric Range	12 KVA	12000 VA
Water Heater	3 KVA	3000 VA
Dishwasher	1.5 KVA	1500 VA
Clothes Dryer	7.2 KVA	7200 VA
Total Existing Loads		**34500 VA**
Demand Factor	34500 VA - 8000 VA = 26500 x 40% =	10600 VA
Addition of A/C Unit	5 KVA	50000 VA
New Total Load (VA)	10600 VA + 8000 VA + 5000 VA =	23600 VA
New Total Load (Amps)	23600 VA / 240V =	**98.33 Amps**
Recommended Service Power Supply		**150 Amps**

7.8 Chapter 7 Review

1. The amount of volts that you get from a single-pole residential breaker is _____ volts.

2. The most common size panels used in residential applications are _____, _____, and _____ amps.

3. Cabinets for electrical panels must be _____ when installed outside.

4. Type-1 panel enclosures provide protection against _____ and _____.

5. New changes to codes require that main lugs and bus bars have _____.

6. A 20/40 type load center has spaces for 20 full size single-pole or 20 _____ breakers.

7. The _____ bar in a subpanel must be isolated using a plastic insulator.

8. _____ circuit breakers are the most common type found in electrical panels.

9. _____ breakers are used for circuits placed in wet areas.

10. _____ breakers help prevent electrical fires.

CHAPTER 7 NOTES:

Electrical Receptacles   8

Electrical receptacles are the closest interaction most people have with electricity in their everyday lives. In this chapter we will outline the types of receptacles and how they are installed.

The most common type of receptacle used in residential areas are duplex receptacles. Each duplex receptacle consists of two outlets which have matching features. Each receptacle has a short contact slot for the hot conductor (black/red) and a long contact slot for the grounding conductor (white), as well as a grounding contact.

Each outlet has a brass and silver terminal to connect conductors. A green terminal is also present for the grounding wire.

<u>Duplex receptacles also have other important components that are listed below:</u>

Mounting Straps - Are used to secure the receptacle to the outlet box.

Connection Tabs - These tabs are centrally located on the receptacles and can be removed depending on the type of installation desired.

Amp and Voltage Rating - Listed on the front of the receptacle for max values allowed.

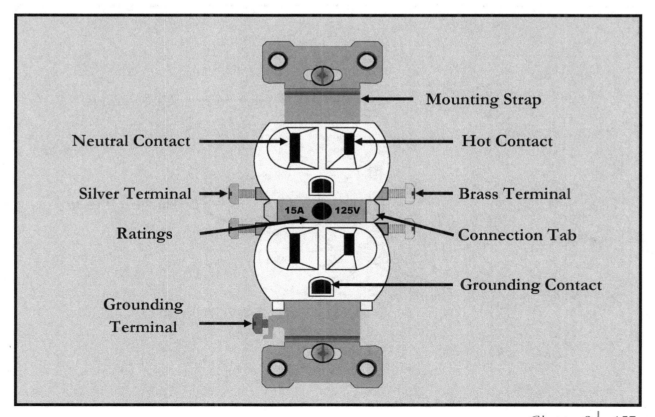

UL Stamp - The UL stamp of approval can be found on the front or back of the receptacles indicating the receptacle has met industry standards.

There are several important features on the back of the receptacle that should be noted:

Strip Gauge - The strip gauge acts as a guide for the removal of the sheath from the conductors. It provides an accurate length of conductor to be stripped for connection to the switch.

Push-in Fittings - Also known as terminal holes, these can be used in certain applications to install conductors instead of using the side terminals.

Conductor Sizing - This designation, found on the center of the receptacle, indicates the max AWG size allowed to be used safely. In most duplex applications the AWG size designation will say #12 or #14.

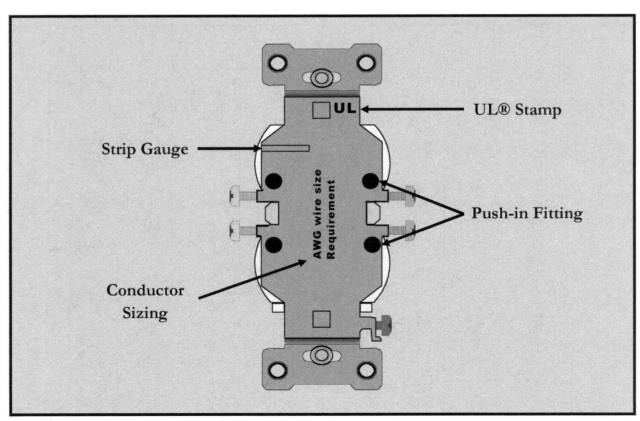

8.1 Installing Receptacles

Even though there are several types of receptacles and switches, they all follow the same basic installation steps with slight variation in the number of wires that you will need to connect. You must first learn how to properly prepare your conductors to be connected securely to receptacles. **These basic guidelines can also be followed for the installation of receptacles and switches.**

Remove around 5/8" of the sheath from the conductors to make the proper connection to the receptacle. This is done with any of the wire strippers discussed earlier. This should be done to all the conductors that will be connected.

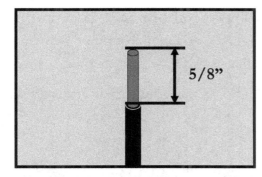

Make a loop with the exposed conductor with the help of needle-nose pliers or some other electrical tool. Make sure the loop can fit around the screw of the receptacle. Do this to all the conductors being connected including the grounding wire.

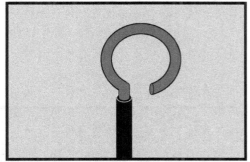

Place the looped conductor around the screw terminal and make sure there is a connection between the screw and the looped conductor. Double check that the correct conductors are connected to the proper terminals.

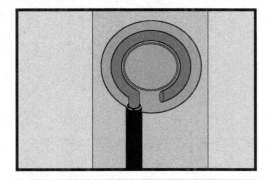

When the looped conductor has been secured around the screw terminal, tighten the screw clockwise until it is snug against the conductor. Give the conductor a light tug to ensure it will not become loose and disconnect. If too much conductor is exposed from when you removed the sheath, then use electrical tape to cover the bare wire.

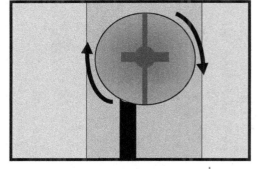

8.2 Duplex Receptacle Installation

Installation of duplex receptacles can differ depending on if you are using a metal or plastic box. The guidelines for installing duplex receptacles with metal boxes are outlined in the *NEC®* handbook. Below you can see the plastic box installation vs. the installation in a metal box.

The metal box uses a pigtail jumper attached to a silver terminal inside the box. This green pigtail jumper is then spliced to the grounding wire and another green pigtail that is attached to the grounding terminal on the duplex receptacle.

The other conductors are attached the same way for both boxes. Attach the white conductor to the silver terminal on the receptacle and the hot (black) conductor to the brass terminal.

You should also note that metal electrical boxes will have some type of mounting device on the knockout to secure the cable.

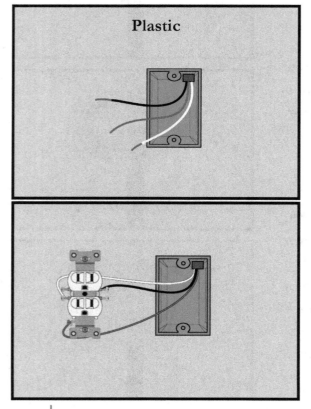

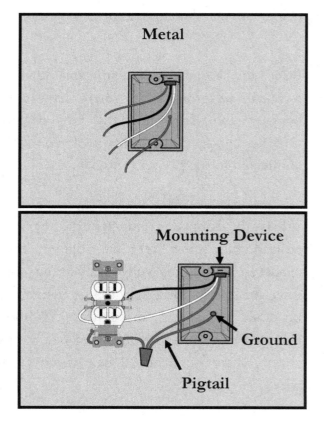

Back-Wire Duplex Installation

The use of the push-in fittings for back-wire installation can only be used under certain guidelines set by the *NEC®*. The *NEC®* states that you may never use back-wire installation if you are continuing a circuit run, such as a multi-wire circuit where there are multiple receptacles on the same circuit. **You may only use the back holes if it is the termination point on the circuit.**

Back-wire installation steps:

- **Strip the white and black conductor** wires according to the strip gauge on the back of the receptacle. The length of the strip gauge gives an accurate length of wire that will enter the holes.

- Insert the **black conductor** into a hole nearest to the brass terminals.

- Insert the **white conductor** into the holes closest to the silver terminals which are opposite of the black push-in holes. Some receptacles will indicate where to place the white wire with a word designation. Some receptacles may also allow you to secure the white and black wires by tightening the terminals.

- Connect the **grounding conductor** to the green grounding screw terminal.

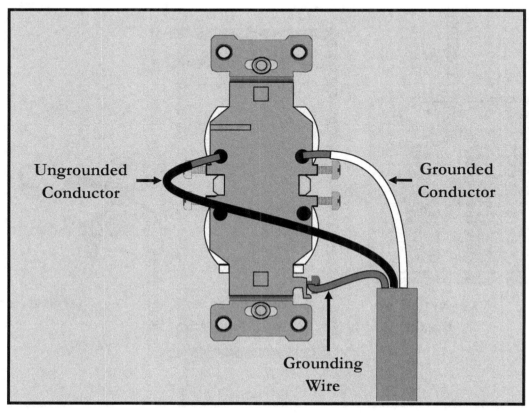

Ungrounded Conductor

Grounded Conductor

Grounding Wire

8.3 Split-Wire Duplex Receptacle Installation

Option 1

This option uses a 2-conductor cable with a grounding wire. Each cable connects half of the receptacles to the overcurrent protection device located on the breaker in the electrical panel. These two cables can be connected to 1 double-pole circuit breaker or 2 single-pole circuit breakers that are attached with a handle tie.

- **Remove the tab** between the brass screw terminals which breaks the connection between each outlet. You do not need to remove the tab on the silver terminal.

- Connect the **white** conductors from the cables to the silver terminals on the receptacle.

- Connect the **black** conductors from the cables to the brass terminals on the receptacle. Make sure the white wire and the black wire from the same cable are connected parallel to each other.

- Connect the **grounding wires** from each cable to the green pigtail attached to the metal box along with a green pigtail connected to the green terminal on the receptacle.

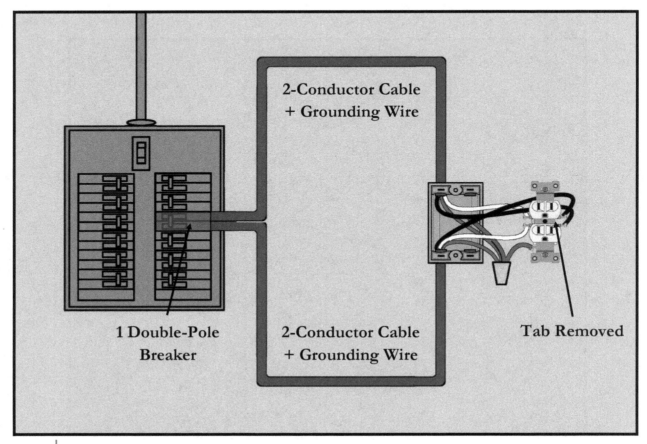

1 Double-Pole Breaker

2-Conductor Cable + Grounding Wire

Tab Removed

Option 2

This options uses one 3-conductor cable with a grounding wire. This cable connects each half of the receptacle to the overcurrent protection device located on the breaker in the electrical panel. The cable must be connected to 1 double-pole circuit breaker.

- **Remove the tab** between the brass screw terminals which breaks the connection between each outlet. You do not need to remove the tab on the silver terminal.

- Connect the **white** conductor from the cable to the silver terminal on the receptacle.

- Connect the **black and red** conductors from the cable to the brass terminals on the receptacle.

- Connect the **grounding wire** from the cable to the green terminal on the receptacle.

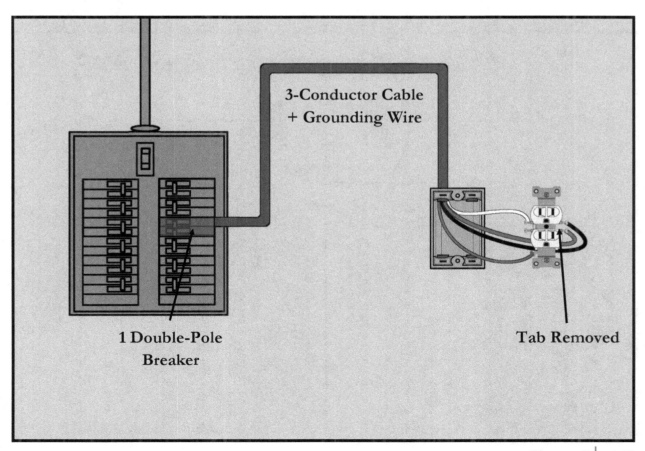

8.4 Range/Dryer Receptacle Installation

Range and dryer receptacles have letter designations on the outlets that are used as a guide for connecting the conductors to the receptacle. These letter designations are:

W - white or grounded circuit conductor

X, Y, Z - the red and black ungrounded (hot) circuit conductors

G - grounding wire connection

If there is no marking, then it is used for hot conductors. **Range cables may have a grounding conductor with no sleeve. However, for dryer applications, a green sleeved grounding wire is needed. These two receptacles also differ in amperage rating. The dryer uses a 30-amp receptacle where the W terminal is in the shape of an "L".**

For range applications, a 50-amp receptacle is used with a straight W terminal instead of the "L" shape.

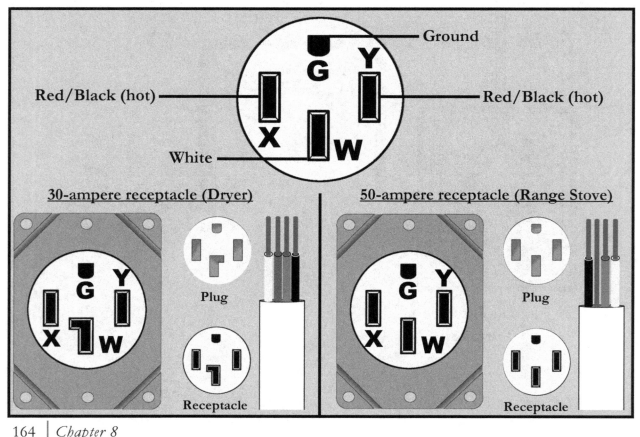

Wiring the range and the dryer follow the same basic wiring principles. However, as discussed before, each appliance uses different amperage receptacles. This will affect the overcurrent protection needed at the breaker and the conductor sizes may change as well. You should follow the regulations for installation set forth by the *NEC®* and the receptacle manufacturer.

The basic wiring diagram for ranges and dryers is shown below:

- Connect the **green** grounding conductor to the rounded "G" terminal located on the receptacle.

- Connect the **white** conductor to the **"W"** terminal on the receptacle.

- Connect the **black** (hot) conductor to the **"Y"** terminal.

- Connect the **red** (hot) conductor to the **"X"** terminal.

- Each cable is a 3-conductor cable with a grounding wire. These should be dedicated branch circuits which run directly to the breaker panel.

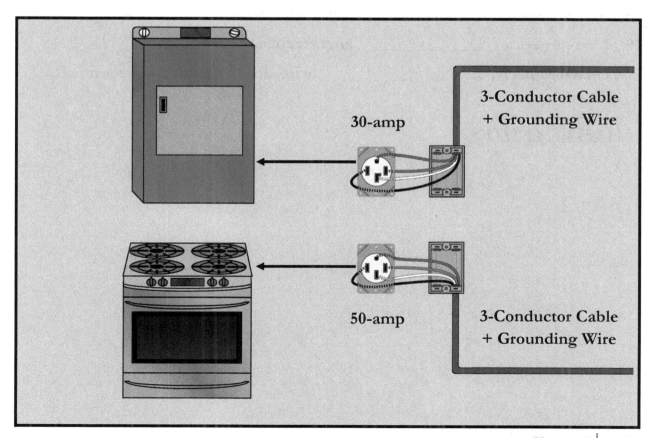

8.5 Chapter 8 Review

1. The most common type of receptacles used in residential areas are _____ receptacles.

2. The conductor must be secured to the screw terminal by rotating the screw in a _____ direction.

3. To install split-wire receptacles you must remove the _____ between the brass screw terminals.

4. X, Y, and Z are used as _____ terminals on 30 and 50-amp receptacles.

5. When installing a receptacle in a metal box, the grounding wire must be _____ to the box.

6. Push-in points in the back of the receptacle are only allowed when it is the _____ point.

7. A dryer receptacle must have a _____ sleeved ground wire.

8. A dryer uses a _____ amp receptacle.

9. A range uses a _____ amp receptacle.

10. On a receptacle, the _____ wire attaches to the brass terminal.

CHAPTER 8 NOTES:

Electrical Switches

9

Electrical switches are devices that create and break a mechanical connection to the circuit. Each switch will have a toggle that can be flipped to make or break this connection.

All switches follow the same wiring principles with slight variations. The figure below shows the basic design of the single-pole switch. **Each switch is stamped or engraved with voltage and amperage ratings along with a UL stamp. The basic design of a switch includes:**

Mounting Holes and Strap - used to secure the switch to the box

Toggle - makes or breaks the connection to supply electrical power

Terminals - connects conductors by securing the wires with tightening screws

The configuration of the screws vary greatly depending on the type of switch.

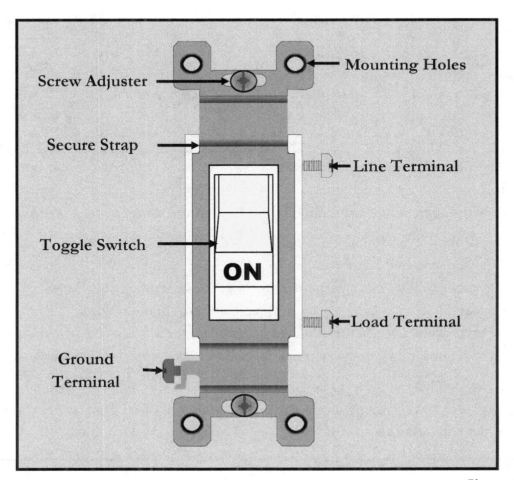

9.1 Switch Types

Although you may encounter very different looking switches with various installations, they are all part of 4 variations of switches. These variations are used when there is an increase in voltage, or when you want to control lighting from several locations on the same line.

<u>There are several types of switches that you will encounter that will have different circuit configurations:</u>

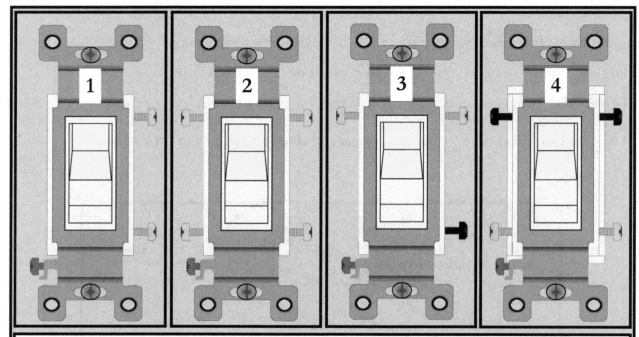

1. **Single-Pole** - These have 3 terminals. Two are brass colored for the line and load conductors, and 1 is green for the grounding wire. These switches control a single device with the possibility of additional devices for the same single source switch.

2. **Double-Pole** - The double-pole switch has 5 terminals. Four are brass colored and used for the line and load conductors, and 1 green screw for the grounding wire. These switches allow for an additional hot wire for controlling 240V devices.

3. **Three-way** - These have 4 terminals. Two are brass colored for the traveler conductors. One is black for alternating load/line, and 1 is green for the grounding conductor. These allow you to control lights or devices from two switches.

4. **Four-way** - The four-way switch has 5 terminals. Two are brass colored and 2 are black for the traveler conductors. One green screw is used for the grounding wire. These switches are used to control lighting from 3 or more locations.

9.2 Switch Variations

Even though there are only 4 major types of switches, there are many variations of these types of switches. Deciding which switch to install is up to the preference of the homeowner. The different styles can add functionality or automation to switches in the residence or workplace.

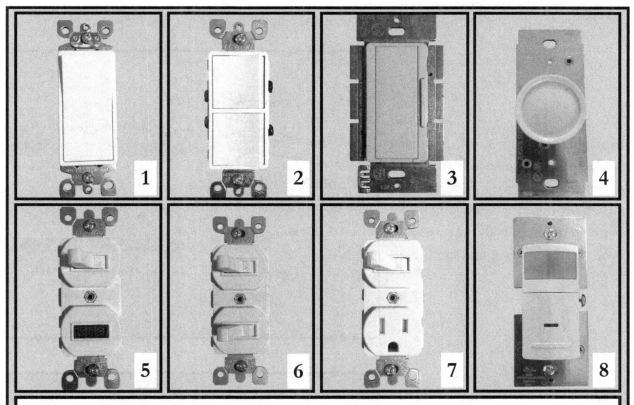

1. **Rocker Switch -** simple flip switch alternative

2. **Double Rocker Switch -** simple double flip switch alternative

3. **Dimmer Switch -** electric one press dimmer switch

4. **Rotary Dimmer Switch -** easy turn dimmer switch for light or fan

5. **Switch with Pilot Light -** has visual indicators of supplied power

6. **Double Switch -** controls two lights or devices

7. **Switch with Receptacle** - controls one lighting device and supplies an outlet

8. **Motion Sensor with Option Switch -** has a motion sensor to activate the lighting fixture

9.3 Single-Pole Switch

When we refer to on and off switches for supplying power to a single light, we are talking about the single-pole switch. The single-pole or throw switch is the most basic of the switches you will encounter, and the most common. Single-pole switches have a single wire considered "hot". Breaking or making this connection with the help of the switch toggle is what powers the device connected to the switch.

When the toggle is switched **ON**, usually in the up position, a spring loaded gate in the switch snaps the internal conductor closed. This closed conductor then links the contact points to allow power to flow between the line and load conductor.

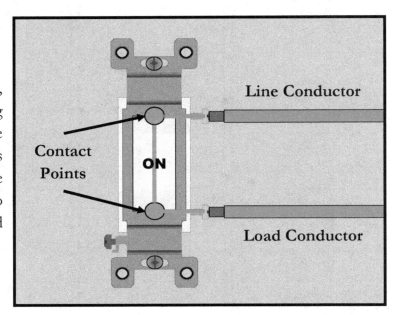

When the toggle is switched **OFF**, usually in the down position, the spring loaded gate moves the internal conductor to break the link between the contact points. When the connection is broken, the flow of electrical power between these points is cut off and prevents power from reaching the light or device.

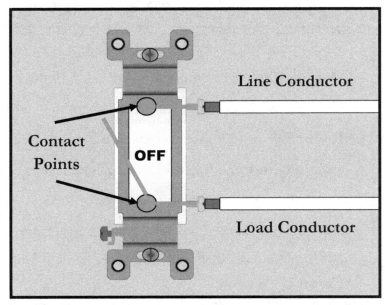

Installing Single-Pole Switches

Single-pole switch installations vary more than any other type of switch. When switches are wired, both the line and the load wires should be designated as "hot". You are allowed to have a white wire used as the "hot wire", but it must be wrapped in black electrical tape to show it is a hot wire.

Switch Leg

The connection is made by taking the black (hot) wire from the power source cable and connecting it to the white conductor from the switch leg cable above the light fixture. The white conductor from the power source cable is connected to the silver light fixture, and the black conductor from the switch leg cable is connected to the brass terminal on the light fixture.

On the switch side of the cable, **the white wire (with black tape) is connected to the line (top) terminal, which is brass. The black wire is connected to the load terminal (bottom) terminal, which is also brass.** The grounding wire which is bare, is connected to the green terminal.

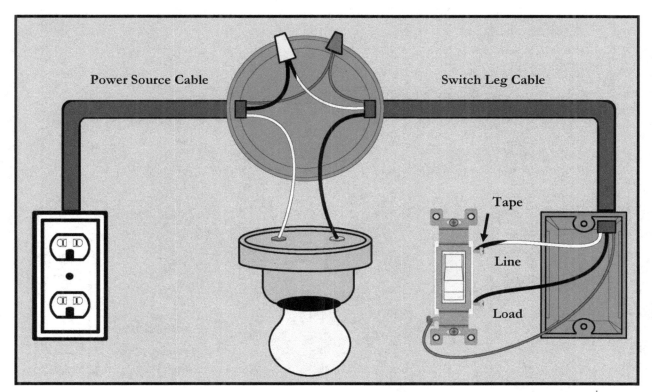

Single-Pole With Power Source Entering at Switch

One cable is run from the outlet which connects the switch to the overcurrent protection device in the panel. The second cable is run from the switch box to the lighting device.

Connecting the Switch

- The black (hot) conductor from the power source cable (outlet) is connected to the line (top) brass terminal of the switch.

- The other black (hot) conductor running from the light switch is run to the load (bottom) brass terminal.

- Connect the two white wires, one from the outlet and the other running to the light, with a wire nut.

- Connect the two bare copper wires from each cable with a pigtail copper wire which is connected to the green grounding terminal.

Connecting the Lighting Fixture

- Connect the white wire to the silver terminal on the light fixture (load terminal).

- Connect the black (hot) wire to the left brass screw on the lighting fixture. The grounding wire is not connected and should be wrapped.

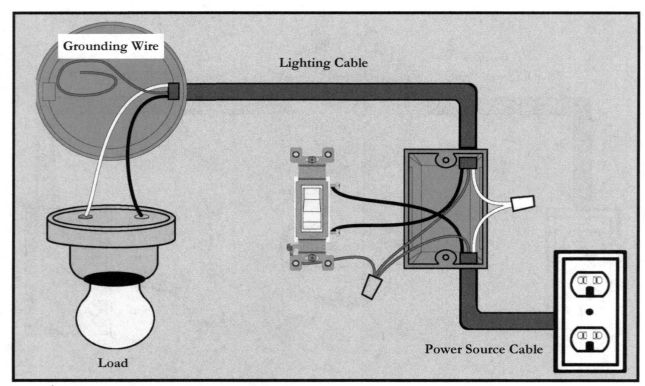

Single-Pole With Two Lighting Sources and Power Source Through Light

Fixture 1

- Connect the black wire from the power source cable to the white wire from cable 1. Make sure to wrap the white wire in black tape to designate it as hot.

- Connect the white wires from the power source and cable 2 to a white pigtail connected to the silver terminal on fixture 1.

- Connect the black (hot) wires from cable 1 and cable 2 with a black pigtail connected to the brass terminal on fixture 1.

- Connect the grounding wires from cable 1, power source cable, and cable 2 together.

Fixture 2

- Connect the black (hot) wire to the brass terminal on fixture 2

- Connect the white wire to the silver terminal on fixture 2. The grounding wire is not connected and should be wrapped.

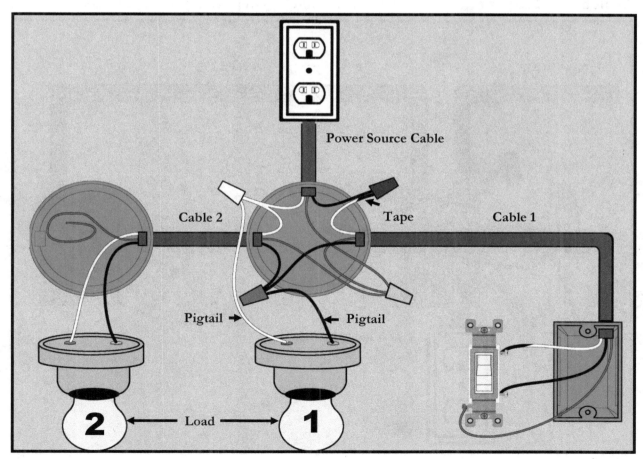

Switched, Split-wire Duplex Receptacle

Determine which outlet you would like for the switch to control and remove the tab on the side of the black (hot) connections on the outlet. It will be between the brass terminals.

Receptacle

- Connect the white wire with black tape from the switch leg cable to the black power source wire and the black wire running to the next receptacle. These wires need to be connected to a black pigtail which is connected to the bottom brass terminal of the receptacle.

- Connect the white wires from the next receptacle and the power source to a white pigtail wire which is run to the top silver terminal on the receptacle.

- Connect the black wire from the switch leg to the top brass terminal on the receptacle.

- Connect the grounding wires from the switch leg, power source cable, and next receptacle cable to a grounding pigtail connected to the green terminal on the receptacle.

Switch connections are the same as the switch leg step from previous page.

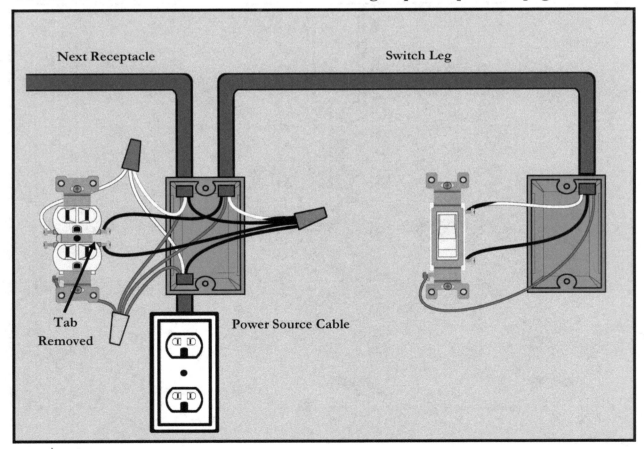

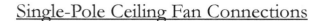
Single-Pole Ceiling Fan Connections

Ceiling fan installations are common in residential installations and can require 1 or 2 single-pole switches. You need 1 switch for a fan without a light and 2 for a fan with a light. Always follow the guidelines outlined earlier for fan installation and box requirements.

The fan motor will contain 4 wires for providing power to the fan and to the light.

Switch Connections

- Connect the black wire from the power source cable to 2 black pigtails. These pigtails are run to the top brass terminals on switch 1 and 2.

- Connect the black wire from cable 1 to the bottom brass terminal on switch 2.

- Connect the red wire from cable 1 to the bottom brass terminal on switch 1.

Fan Connections

- Connect the red wire from cable 1 to the blue wire and the grounding wire to the green wire connected to the fan. Connect the white wire from cable 1 to the white fan wire.

- Connect the two white wires together and the black wires together.

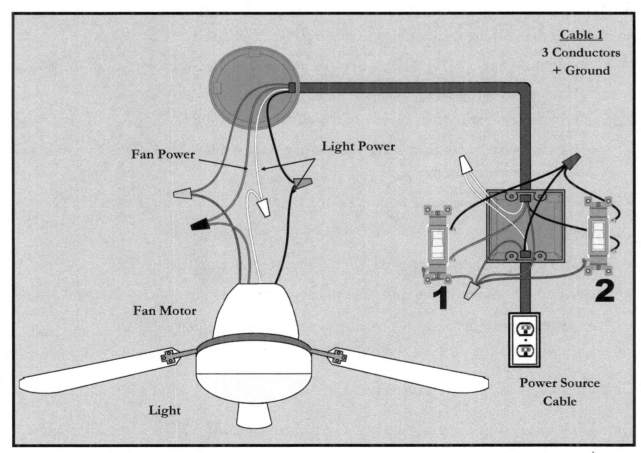

Two Single-Pole Switches Connected to Two Fixtures With Power Through Fixture 1

When installing 2 single-pole switches to two fixtures, an additional wire (red) is added to account for the added switch. Therefore, a 3-conductor cable with a grounding wire must be used from the switches to the first fixture, then a 2-conductor cable with grounding wire from fixture 1 to 2.

Switch Connections

- Connect the red wire from cable 1 to the bottom brass terminal of switch 1.

- Connect the black pigtails linked to the top brass receptacles of switch 1 and 2 to the white wire from cable 1. Cover the white wire in black tape to designate it as hot.

- Connect the black wire from cable 1 to the bottom brass terminal on switch 2.

- Connect the grounding wire pigtails connected to the green terminals from switch 1 and 2 to the grounding wire of cable 1.

Fixture 1

- Connect the red wire from cable 1 to the brass terminal on fixture 1.

- Connect the white wire with black tape from cable 1 to the black wire from the power source.

- Connect the black wire from cable 1 to the black wire of cable 2.

- Connect the white wires from the power source and cable 2 to a white pigtail connected to the silver terminal on fixture 1.

- Connect the grounding wires from the cable 1, cable 2, and the power source cable.

Fixture 2

- Connect the white wire from cable 2 to the silver terminal on fixture 2.

- Connect the black wire from cable 2 to the brass terminal in fixture 2.

- The grounding conductor is not connected and wrapped securely.

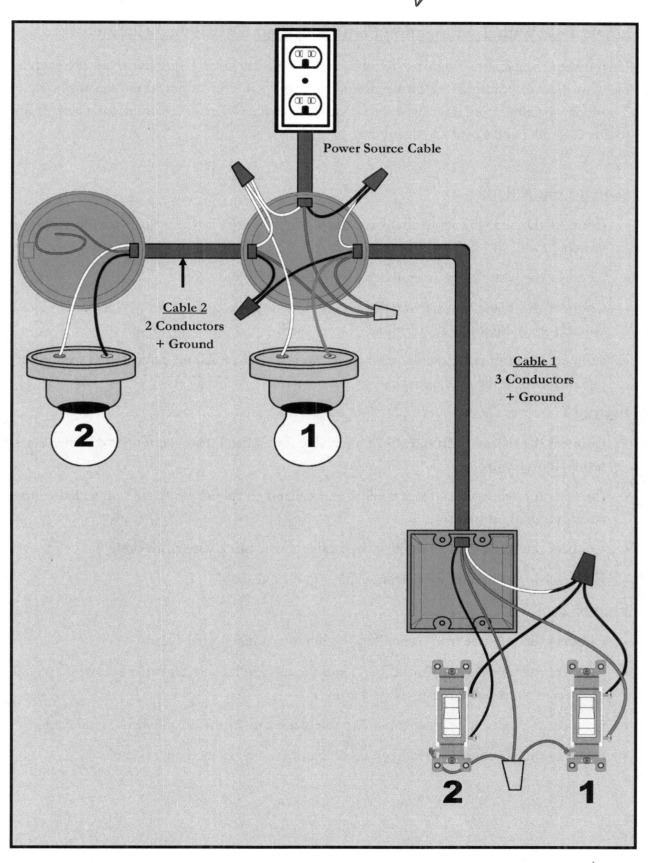

Power Source Cable

Cable 2
2 Conductors
+ Ground

Cable 1
3 Conductors
+ Ground

Single-Pole With Two Lighting Fixtures and an Unswitched Circuit

If you want to install two lighting fixtures with the power source running from the switch, but you wish to continue the circuit to another receptacle or outlet, then you must use a 3-conductor cable from the switch to fixture 1 and 2. Then a 2-conductor cable from fixture 2 to the next receptacle or fixture.

Switch Connections

- Connect the red wire from the 3-conductor cable to the bottom brass terminal on the switch.

- Connect the white wires from the power source cable and cable 1.

- Connect the black wires from the power source cable and cable 1 to a black pigtail connected to the top brass terminal of the switch.

- Connect the grounding wires from the power source cable and cable 1 to a grounding pigtail linked to the green terminal on the switch.

Fixture 1

- Connect the red wires from cable 1 and cable 2 to a black pigtail connected to the brass terminal on fixture 1.

- Connect the white wires from cable 1 and cable 2 to the white pigtail connected to the silver terminal on fixture 1.

- Connect the black wire from the switch cable to the black wire from cable 2.

- Connect the grounding wires from cable 1 and cable 2.

Fixture 2

- Connect the red wire from cable 2 to the brass terminal on the fixture.

- Connect the white wires from cable 2 and the unswitched circuit cable to a white pigtail connected to the silver terminal on fixture 2.

- Connect the black wire from cable 2 to the black wire in the unswitched circuit cable.

- Connect the grounding wires from cable 2 and the unswitched circuit cable.

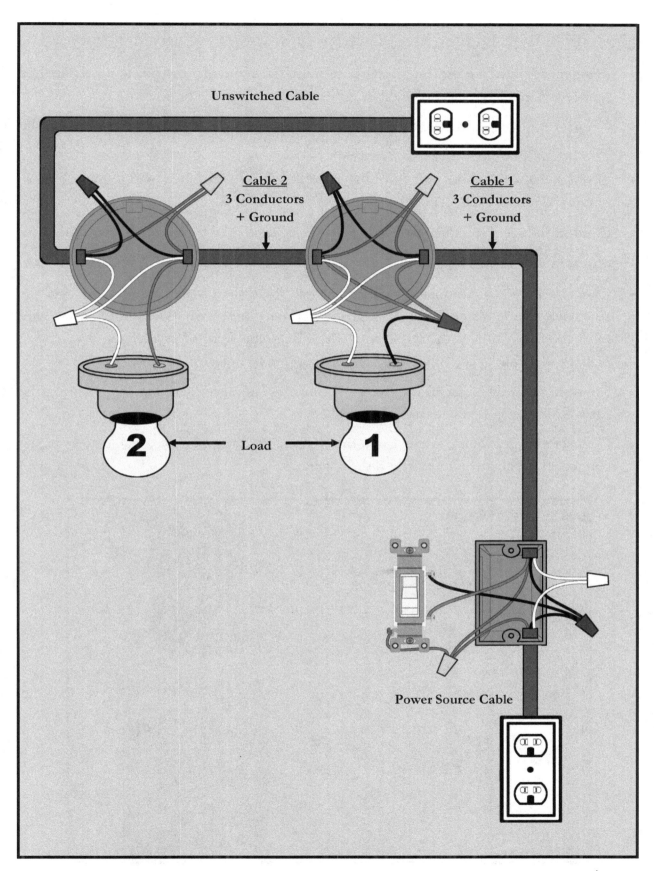

Unswitched Cable

Cable 2
3 Conductors
+ Ground

Cable 1
3 Conductors
+ Ground

2 ← Load → 1

Power Source Cable

Single-Pole With Duplex Receptacle for Dishwasher / Garbage Disposal

- Remove the tab on the brass screw terminal side on the receptacle to make it a split-receptacle. This allows for the separate garbage disposal connection.

- Connect the black wire from the switch leg cable to the bottom brass terminal on the switch.

- Connect the white conductor from the switch leg cable with black tape to the top brass terminal on the switch.

- Connect the green conductor from the switch leg cable to the grounding terminal.

Receptacle

- Connect the black conductor from the switch leg cable to the brass terminal on the receptacle that is intended for the garbage disposal connection. Connect the black wire from the SEP cable to the white wire with black tape from the switch leg and a black pigtail connected to the other brass terminal on the receptacle.

- Connect the white wire from the SEP cable to the silver terminal opposite of the garbage disposal black conductor connection.

- Connect the grounding wires from the SEP cable and switch leg cable to a pigtail connected to the green terminal on the receptacle.

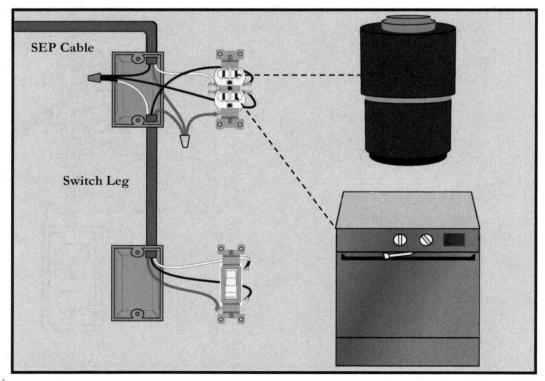

9.4 Three-Way Switches

The three-way switch is given its name based on the number of terminals it has (3). There are a few things that make three-way switches different than your normal single-pole switches including:

There are two brass terminals on the switch. These terminals are "traveler" terminals. There is a black "common" terminal on the switch and is sometimes referred to as the hinge.

Since three-way switches act together with other three-way switches, there is no ON or OFF setting for the toggle. Three-way switches are meant to be used with other three-way switches and should always be used in pairs and not alone.

Three-way switches have three internal contacts where connections are made. Whether the toggle is up or down, there is always a connection between the load conductor connected to the dark terminal and one of the traveler conductors. A connection is always made in three-way switches.

The pictures below show how three-way switches function. **The wires that are orange represent the electrical load, and not the actual color of the wires that are connected to the terminals.**

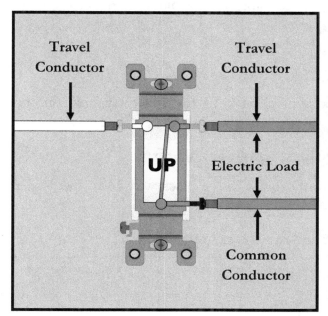
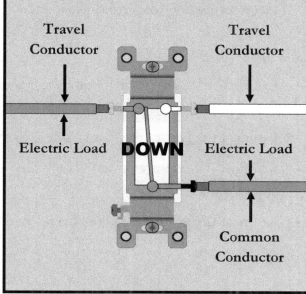

Three-Way Switch With Power Source Through Light Fixture

Switch 1

- Connect the red wire from cable 1 to the top left brass terminal on switch 1.

- Connect the black wire from cable 1 to the top right brass terminal on switch 1.

- Connect the white wire with tape from cable 1 to the bottom dark common terminal on switch 1.

- Connect the bare copper grounding wire from cable 1 to the green grounding terminal on switch 1.

Switch 2

- Connect the red wire from cable 1 to the top right brass terminal on switch 2.

- Connect the black wire from cable 1 to the top left brass terminal on switch 2.

- Connect the black wire from cable 2 to the bottom dark common terminal on switch 2

- Connect the white wire from cable 1 and the white wire from cable 2 with black tape to designate the wires as hot.

- Connect the grounding wires from cable 1 and cable 2 to a pigtail connected to the green grounding terminal on switch 2.

Fixture

- Connect the white wire from the power source cable to the silver terminal on the fixture.

- Connect the black wire from cable 2 to the brass terminal on the fixture.

- Connect the white wire with black tape from cable 2 to the black wire from the power source cable.

- Connect the grounding wires from cable 2 and the power source cable.

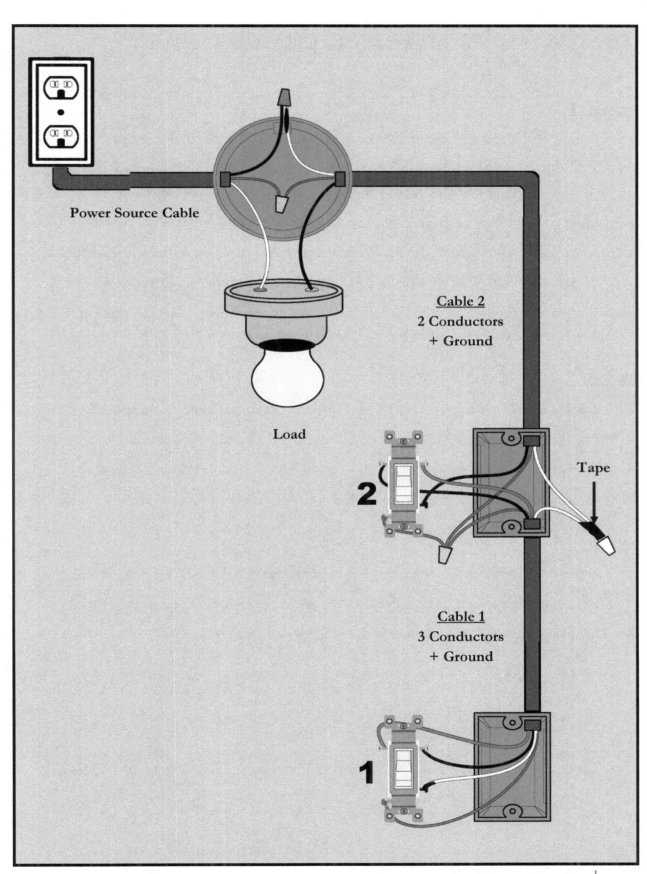

Power Source Cable

Cable 2
2 Conductors
+ Ground

Load

Tape

2

Cable 1
3 Conductors
+ Ground

1

Three-Way Switch With Power Source Through First Switch

Switch 1

- Connect the black wire from the power source cable to the dark common terminal on switch 1.

- Connect the red wire from cable 1 to the top left brass terminal on switch 1.

- Connect the black wire from cable 1 to the top right brass terminal on switch 1.

- Connect the white wire from the power source cable to the white wire from cable 2.

- Connect the white wire with black tape from cable 1 to the black wire from cable 2.

- Connect the grounding wires from the power source cable, cable 1, and cable 2 to a pigtail connected to the green grounding terminal on switch 1.

Switch 2

- Connect the red wire from cable 1 to the top left brass terminal on switch 2.

- Connect the black wire from cable 1 to the top right brass terminal on switch 2.

- Connect the white wire with black tape to the dark common terminal on switch 2.

- Connect the grounding wire from cable 1 to the green grounding terminal on switch 2.

Fixture

- Connect the black wire from cable 2 to the brass terminal on the lighting fixture.

- Connect the white wire from cable 2 to the silver terminal on the lighting fixture.

- The grounding wire is not connected and is looped and secured.

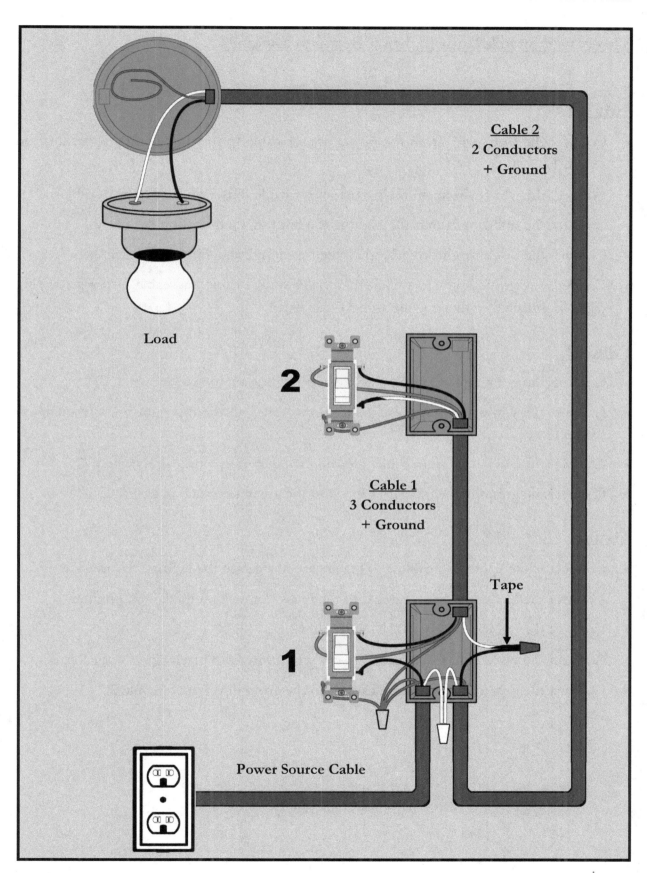

Cable 2
2 Conductors
+ Ground

Load

2

Cable 1
3 Conductors
+ Ground

Tape

1

Power Source Cable

Three-Way Switch With Lighting Between Switches

Switch 1

- Connect the black wire from the power source cable to the dark common terminal on switch 1.

- Connect the white wire from the power source cable to the white wire from cable 1.

- Connect the red wire from cable 1 to the top left brass terminal on switch 1.

- Connect the black wire from cable 1 to the top right brass terminal on switch 1.

- Connect the grounding wires from the power source cable and cable 1 to a grounding pigtail connected to the green terminal on switch 1.

Switch 2

- Connect the red wire from cable 2 to the top left brass terminal on switch 2.

- Connect the white wire with black tape from cable 2 to the top right brass terminal on switch 2.

- Connect the black wire from cable 2 to the dark common terminal on switch 2.

- Connect the grounding wire from cable 2 to the green terminal on switch 2.

Fixture

- Connect white wire from cable 1 to the silver terminal on the lighting fixture.

- Connect the black wire from cable 2 to the brass terminal on the lighting fixture.

- Connect the red wires from cable 1 and cable 2.

- Connect the black wire from cable 1 to the white wire with black tape from cable 2.

- Connect the grounding wire from cable 1 to the grounding wire to cable 2.

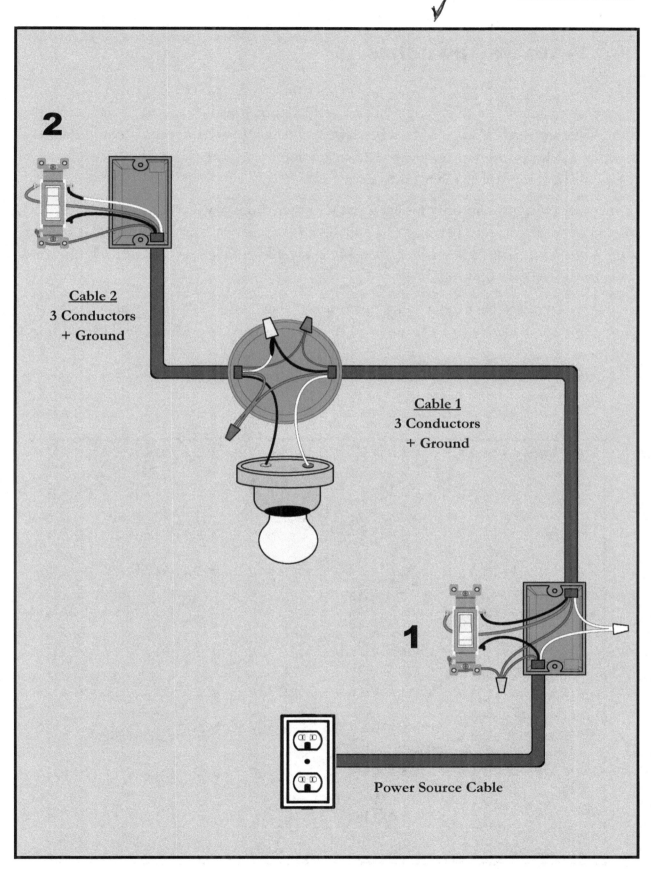

2

Cable 2
3 Conductors
+ Ground

Cable 1
3 Conductors
+ Ground

1

Power Source Cable

9.5 Four-Way Switches

Four-way switches, like three-way switches, are given their name because of the number of terminals they have. Four-way switches have a total of 4 travel terminals comprised of 2 dark and 2 brass colored terminals. There are some variations with color for the travel terminals so always check the manufacturer's guides.

Four-way switches can only be used with 2 other three-way switches. In any situation, the number of four-way switches can only be a maximum of half of the number of three way switches installed. Four-way switches must be installed in between 2 three-way switches, never on the outside.

Four-way switches have no common terminal. There is always a connection with the four-way switch and they can be installed multiple ways. They have no marking for on or off status on the toggle.

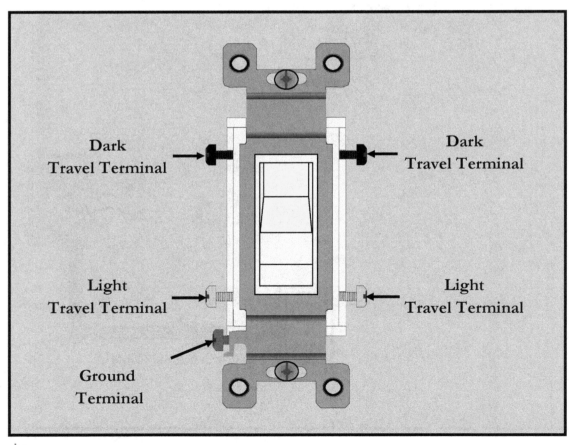

Like all switches, four-way switches change their connections through a series of internal contacts with connecting poles. What makes four-way switches unique is that they posses 2 internal poles for connections.

Because they possess multiple poles, four-way switches can come made with vertical or horizontal poles. These two types designate whether the travel direction of the circuit is vertical or horizontal.

Horizontal

The top two pictures below show how the horizontal connection is made between poles. When the toggle is flipped the connection creates a diagonal pattern.

Vertical

The bottom two pictures below show how the vertical connection is made between poles. Like the horizontal four-way, when the toggle is flipped, the connection creates a diagonal pattern.

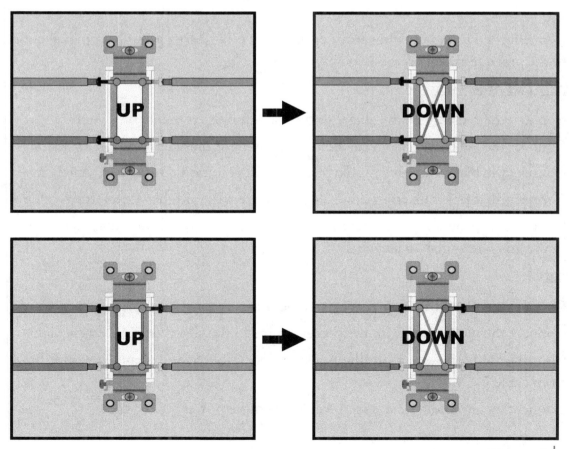

Four-Way Switch Installed in a Vertical Configuration

Switch 1 (Three -Way)

- Connect the red wire from cable 1 to the top left brass terminal on switch 1.
- Connect the black wire from cable 1 to the dark common terminal on switch 1.
- Connect the white wire from cable 1 to the top right brass terminal on switch 1.
- Connect the grounding wire from cable 1 to the green terminal on switch 1.

Four-Way switch

- Connect the white wire from cable 1 to the bottom right brass travel terminal on the four-way switch.
- Connect the red wire from cable 1 to the bottom left brass travel terminal on the four-way switch.
- Connect the black wire from cable 1 to the black wire from cable 2.
- Connect red wire from cable 2 to the top left dark travel terminal on the four-way switch.
- Connect the white wire from cable 2 to the top right dark travel terminal on the four-way switch.
- Connect the grounding wires from cable 1 and cable 2 to a grounding pigtail connected to the green terminal on the four-way switch.

Switch 2 (Three -Way)

- Connect the red wire from cable 2 to the top left brass terminal on switch 2.
- Connect the white wire from cable 2 to the top right brass terminal on switch 2.
- Connect the black wire from cable 3 to the dark common terminal on switch 2.
- Connect the black wire from cable 2 to the white wire with black tape from cable 3.
- Connect the grounding wires from cable 2 and cable 3 to a grounding pigtail connected to the green terminal on switch 2.

Fixture

- Connect the black wire from cable 3 to the brass terminal on the light fixture.
- Connect the white wire from the power source to the silver terminal on the light.
- Connect the black wire from the power source cable to the white wire with black tape from cable 2.
- Connect the grounding wires from cable 3 and the power source cable.

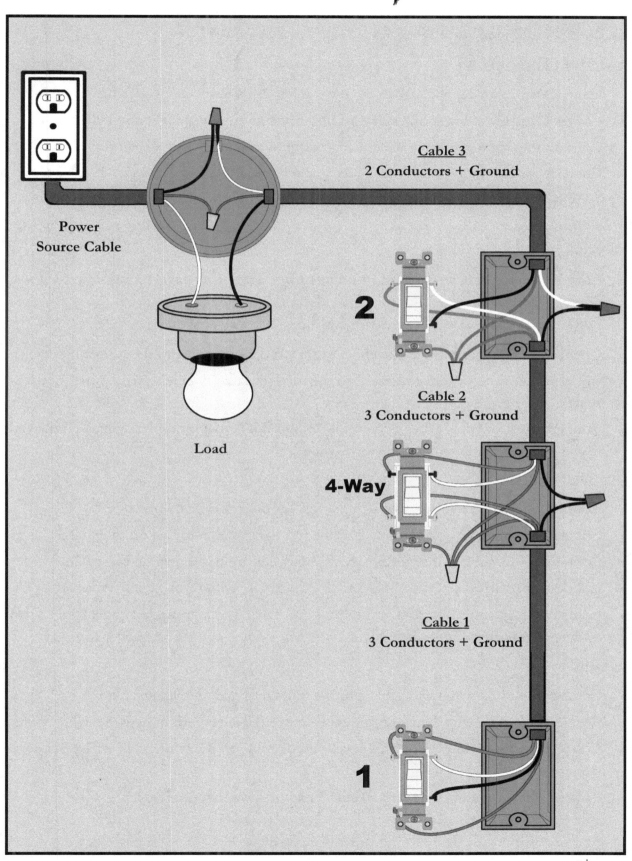

Power
Source Cable

Load

Cable 3
2 Conductors + Ground

2

Cable 2
3 Conductors + Ground

4-Way

Cable 1
3 Conductors + Ground

1

Four-Way Switch Installed in a Horizontal Configuration

Switch 1 (Three -Way)

- Connect the red wire from cable 1 to the top left brass terminal on switch 1.
- Connect the black wire from cable 1 to the dark common terminal on switch 1.
- Connect the white wire from cable 1 to the top right brass terminal on switch 1.
- Connect the grounding wire from cable 1 to the green terminal on switch 1.

Four-Way switch

- Connect the white wire from cable 1 to the top right brass travel terminal on the four-way switch.
- Connect the red wire from cable 1 to the bottom right brass travel terminal on the four-way switch.
- Connect the black wire from cable 1 to the black wire from cable 2.
- Connect red wire from cable 2 to the bottom left dark terminal on the four-way switch.
- Connect the white wire from cable 2 to the top left dark travel terminal on the four-way switch.
- Connect the grounding wires from cable 1 and cable 2 to a grounding pigtail connected to the green terminal on the four-way switch.

Switch 2 (Three -Way)

- Connect the red wire from cable 2 to the top left brass terminal on switch 2.
- Connect the white wire from cable 2 to the top right brass terminal on switch 2.
- Connect the black wire from cable 3 to the dark common terminal on switch 2.
- Connect the black wire from cable 2 to the white wire with black tape from cable 3.
- Connect the grounding wires from cable 2 and cable 3 to a grounding pigtail connected to the green terminal on switch 2.

Fixture

- Connect the black wire from cable 3 to the brass terminal on the light fixture.
- Connect the white wire from the power source to the silver terminal on the light fixture.
- Connect the black wire from the power source cable to the white wire with black tape from cable 2.
- Connect the grounding wires from cable 3 and the power source cable.

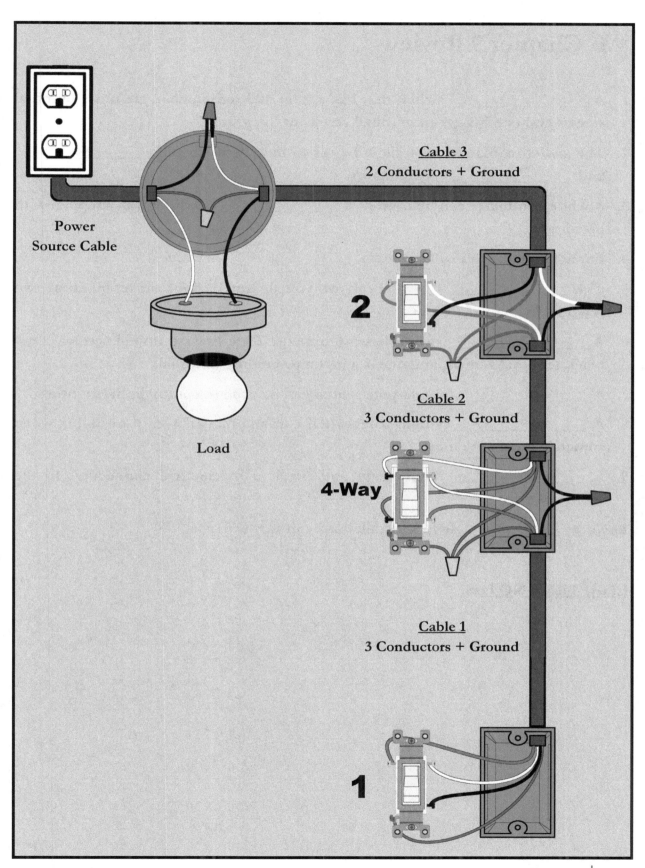

Power
Source Cable

Cable 3
2 Conductors + Ground

2

Load

Cable 2
3 Conductors + Ground

4-Way

Cable 1
3 Conductors + Ground

1

9.6 Chapter 9 Review

1. A _____ switch that has a total of 3 terminals, 2 are brass terminal screws and one is a green terminal screw for the ground.

2. The switch contact points allow power to flow between the _____ and _____ conductors.

3. A white conductor can be used as a _____ wire if it is wrapped with black tape.

4. For a fan/light you will have two _____ switches.

5. The _____ switch allows you to turn a light on or off from two separate locations.

6. A _____ switch has 5 terminals, 2 are brass terminal screws, 2 are black terminal screws, and one is a green grounding terminal.

7. A _____ single-pole switch can be used to control light brightness.

8. A _____ single-pole switch can automatically turn on lights when someone enters the room.

9. A _____ can create and break a mechanical connection to the electrical circuit.

10. _____ switch installations can vary the most.

CHAPTER 9 NOTES:

Grounding and Bonding   10

Proper grounding and bonding is one of the most important steps to making an electrical system safe. Proper grounding also ensures that the system will run properly and protect all of the noncurrent carrying parts of the structure. The *NEC®* code requires that all electrical systems be integrated with electrical grounding systems. The grounding system design can vary in many ways depending on its installation location.

The grounding system performs many important functions, including:

1. Limits high voltages from devices, line surges, and any unintentional contact with separated voltage systems.

2. Controls the voltage transferred to the ground through conductive materials.

3. Stabilizes the voltage transferred to the ground during normal operations.

4. Creates an efficient path for any fault current to ease the operation of overcurrent devices in case of protection failure.

10.1 Grounding Electrodes and Conductors

The grounding electrode conductor connects the grounding electrode to both the equipment grounding conductor and the neutral of the circuit at the service. **The *NEC®* sets guidelines for installation and protection of grounding electrode conductors. The *NEC®* requires that these conductors must be unbroken without the use of bolted splices from the grounding electrode to the neutral connection.**

Aluminum or copper-clad aluminum grounding conductors cannot be used in direct contact with masonry, on the ground, or where it could be exposed to corrosive substances. If these types of conductors are to be buried underground, then **they must be placed at least 18" below the surface if they are not encased in cement.**

Grounding electrodes are normally imbedded underground or in concrete below the surface. This is done to maintain the ground potential for connected conductors in order to effectively dissipate electrical flow into the earth.

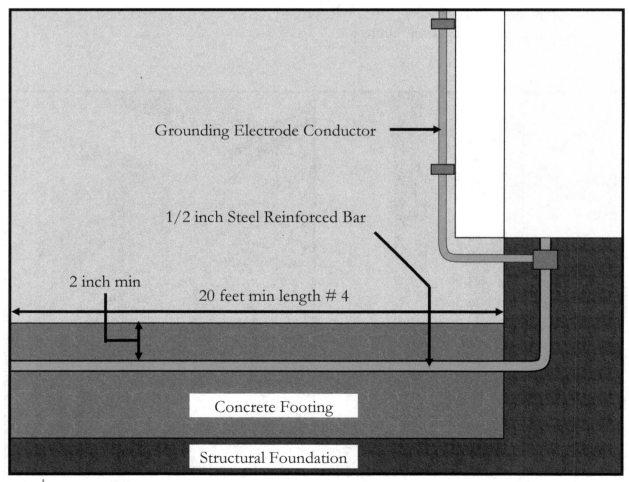

Grounding Electrode Conductor

1/2 inch Steel Reinforced Bar

2 inch min

20 feet min length # 4

Concrete Footing

Structural Foundation

UFER Grounds

UFER type grounds are the preferred type of grounding system used where it is permitted by local codes. The *NEC®* states that UFER ground installations require at least 20 feet of bare copper wire that is no smaller than #4. This wire must be encased by at least 2 inches of concrete and located in a concrete foundation that is in direct contact with the earth.

Other types of grounding systems include driven pipe, driven rod, and buried plate:

- **Driven Pipe:** Galvanized pipe that has a trade size of 3/4" x at least 10 feet driven to a minimum depth of 10 feet.

- **Driven Rod**: Steel or iron rods with a minimum outside diameter of 5/8" x 8 feet. Copper clad rods can also be used with a minimum outside diameter of 1/2" x 8 feet driven to at least 8 feet.

- **Buried Plate:** Iron or steel plates that are at least 0.25" thick. Copper plates can be used that are at least 0.06" thick. These plates must have 2 sq. ft. of surface area in contact with the ground and be buried at least 30" below surface.

Buried plates are not as common as the other types of grounding systems. However, they are ideal in areas where you are not permitted to use grounding rods, pipes, or grounding rings.

Driven rods can be angled depending on the rock bed depth below the surface or if there is a maximum allowed buried depth. Vertical buried rods must be buried to a depth of 8 feet while horizontal rods must be completely buried at least 30 inches below surface.

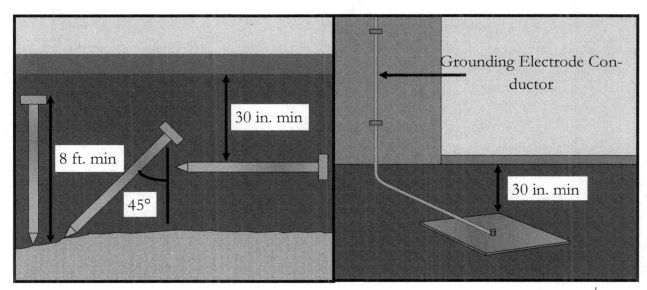

***NEC®* article 250 states the resistance between electrodes and grounds cannot exceed 25 ohms.** In situations where the 25 ohms limit is exceeded, then two or more electrodes should be connected in tandem in a parallel position. These rods should be a minimum of 6 feet apart.

Metal structure framing can also be used to create the grounding electrode system. This usually applies to commercial or industrial structures with metal framing, but some residential structures may also have metal framing. The grounding conductor is secured to the metal beam similar to how it is attached to an underground metal plate.

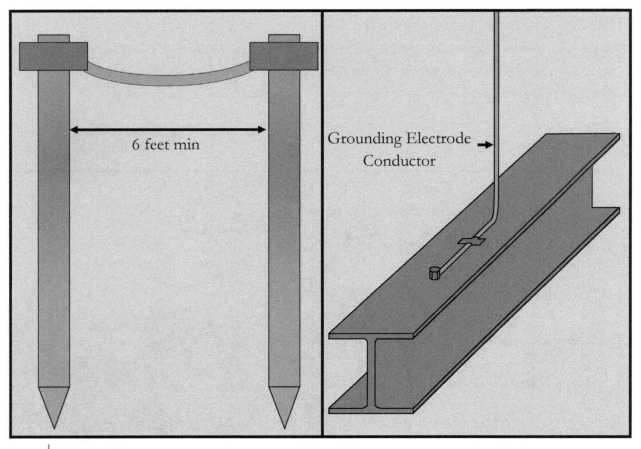

Ground Ring

A ring grounding system is typically used when the use of the other grounding systems is not permitted. These alternative grounding systems are installed at the base of the entire building's grounding structure at **a minimum length of 20 feet exposed to ground. The *NEC®* also states that this system must use at least #2 AWG wire and be buried to a depth of at least 2 1/2 feet.**

Ring grounds should be installed past the building's drip line to prevent any potential corrosion of the ring's metal. You may come into contact with halo ring ground systems. These systems instead of being installed below the structure, are installed near the top of the building. These halo ground systems are connected to the main structure grounding system or an underground ring ground system. These are often used where they can be connected to metal framing structures.

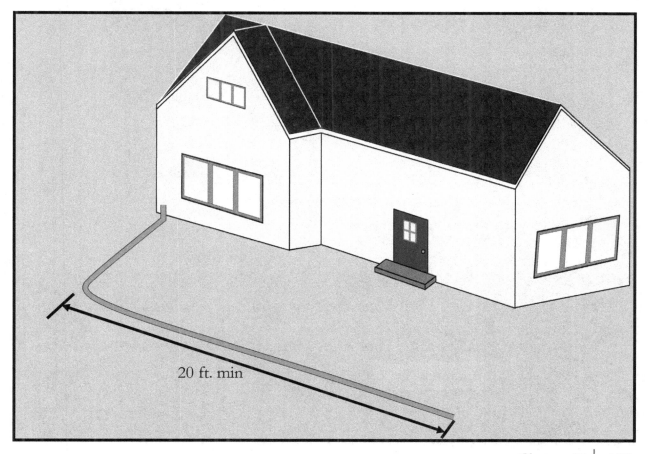

20 ft. min

Metal Underground Water Pipes

In the presence of underground water pipe systems, electrode grounding systems should be bonded directly to the water pipe system through a bonding jumper. The bonding jumper is then connected to the neutral bus bar in the service entrance panel.

The connection to the water pipe requires that it be made within the first 5 feet of it entering the structure. For the water pipe to be considered as a part of the electrical grounding system, it must be in direct contact with at least 10 feet of earth. This ensures that any electrical flow traveling to the water pipe will be properly discharged to the ground.

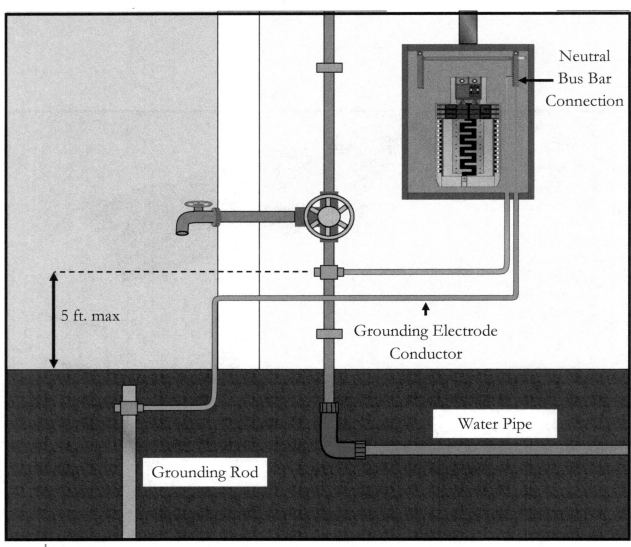

10.2 Grounding Conductor Sizing

The *NEC®* provides guidelines for the minimum grounding electrode conductor sizes needed based on the service entrance conductor sizing and the material used. **These guidelines found in *NEC®* table 250-66 should be followed when designing the grounding system.**

NEC® Table 250-66 Minimum Grounding Electrode Conductor Sizing			
Size of Largest Service Entrance Conductor		Size of Grounding Electrode Conductor	
Copper	Aluminum or Copper-clad Aluminum	Copper	Aluminum or Copper-clad Aluminum
#2 or smaller	1/0 or smaller	8	6
1-1/0	2/0 - 3/0	6	4
2/0 - 3/0	4/0 - 250 kcm	4	2
Over 3/0 to 350 kcm	Over 250 kcm to 500 kcm	2	1/0
Over 350 kcm to 600 kcm	Over 500 to 900 kcm	1/0	3/0
Over 600 kcm to 1100 kcm	Over 900 kcm to 1750 kcm	2/0	4/0
Over 1100 kcm	Over 1750 kcm	3/0	250 kcm

10.3 Bonding

Bonding is performed to ensure that any fault current produced at the device or receptacle is properly conducted through the electrode grounding system. In order to ground equipment, grounding conductors are run from the device to the electrical panel.

The grounding conductor entering the panel is connected to the grounding bus bar on an open terminal. There should be ample room to route the grounding and neutral conductors through the panel space.

This process should be done for all receptacles, devices, and appliances being connected to the panel. Ensure that the grounding conductors do not come into contact with live parts of the panel.

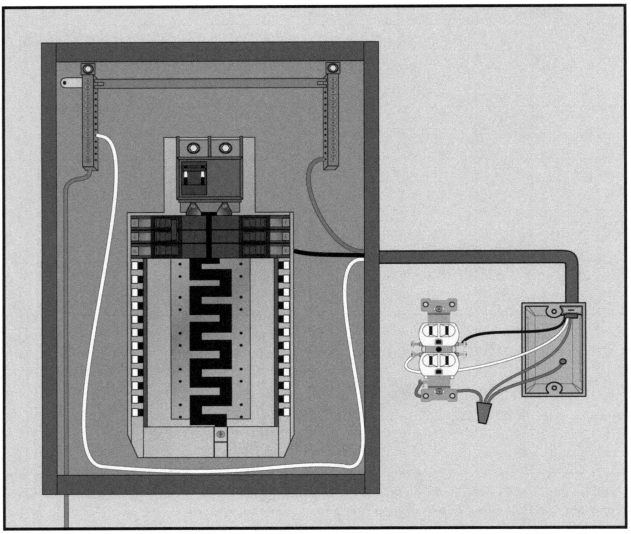

10.4 Fault Current Event

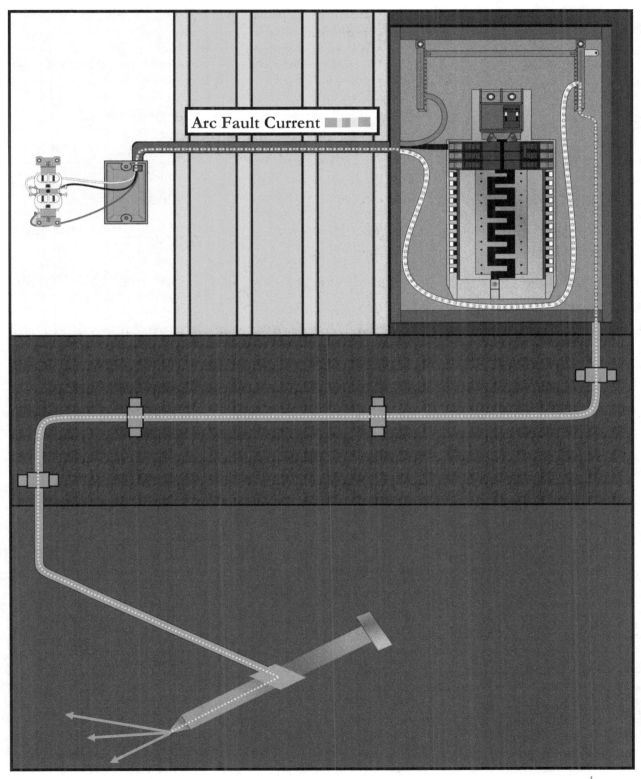

10.5 Chapter 10 Review

1. Proper _____ and _____ is one of the most important steps to making an electrical system safe.

2. Grounding _____ are normally imbedded in concrete or underground.

3. _____ grounds require copper wire no smaller than #4 AWG.

4. A steel or iron rod of 5/8 in. OD must be driven _____ feet into the ground.

5. Grounding electrodes can not exceed _____ ohms.

6. Ground rings must be buried a minimum of _____ feet and must be at least _____ feet long.

7. Grounding electrode conductor sizing can be found in *NEC®* table _____ .

8. _____ will ensure that any fault current is conducted through the electrode grounding system.

9. A grounding conductor entering a panel should be connected to _____ .

10. What size grounding electrode conductor do I need if service conductors are 2/0-3/0? _____ .

CHAPTER 10 NOTES:

GFCI and AFCI

11

GFCI and AFCI protected devices are used to protect people from electrical shock or harm, they do not protect electrical systems. Electrical systems are protected by normal circuit breakers and grounding systems, while GFCI and AFCI receptacles, devices, and breakers are used to shield anyone who may come in contact with them.

In this chapter we will cover the use of GFCI and AFCI in the following sections:

1. How GFCIs Work

2. Types of GFCIs

3. Areas Requiring GFCI

4. How AFCIs Work

5. Types of AFCIs

6. Areas Requiring AFCI

7. Installing GFCI or AFCI

8. Testing GFCI or AFCI

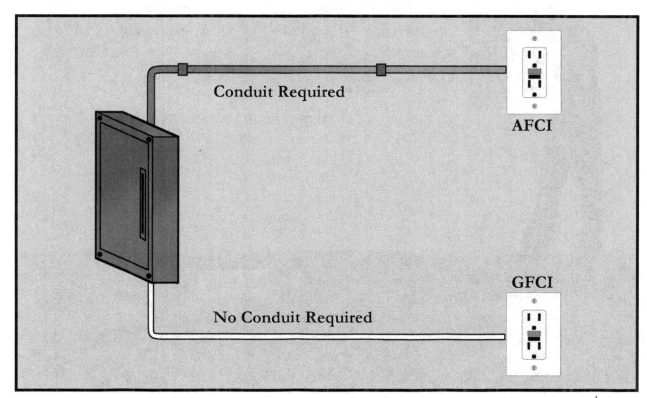

11.1 How GFCIs Work

Ground Fault Circuit Interrupters monitor the current flowing from the hot to the neutral conductors. If there is an imbalance in the flow of the current, it trips either the GFCI circuit breaker, GFCI receptacle, or portable GFCI. **GFCIs can be designed to detect as little as 4 to 5 milliamps of imbalance.**

For example, if you are using a power tool outside in the rain and you are standing on the ground, then you become a grounding path. If a hot wire inside the power tool becomes wet it could create an electrical flow path from the tool, through your body, and into the ground which could cause serious injury or death.

To prevent this from happening, **GFCIs realize that electrical flow is traveling through you because of the change of flow from hot to neutral.** As soon as this realization occurs, the GFCIs can cut the flow of electrical power. These GFCIs can react as fast as 1/13th of a second.

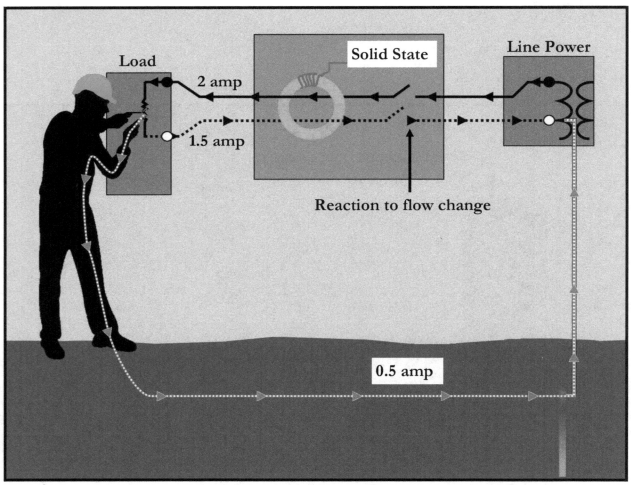

11.2 Types of GFCIs

There are three main types of GFCIs used in homes. The GFCI receptacle, GFCI circuit breaker, and the portable GFCI.

GFCI Receptacle - The most common type of GFCI integration used in residential electrical systems. These receptacles are installed in place of normal receptacles and act as normal duplex receptacles. The functional difference is that GFCI receptacles are equipped with reset and test buttons which allow for GFCI outlets to be tested after they have been tripped. GFCIs can be installed to protect a single outlet or all outlets on its circuit.

GFCI Circuit Breaker - Is installed to protect an entire circuit at the electrical panel. These breakers are installed in the same spaces as normal circuit breakers. These GFCI breakers can be 15 to 60 amps and provide overcurrent protection and GFCI protection to each outlet on the circuit.

Portable GFCI - These are convenient GFCI protection devices to use while working on mobile applications. These GFCIs are often found on worksites, and are used by contractors and electrical workers while in the field. Portable GFCI devices can be utilized while using power tools on the jobsite. They are plugged into normal outlets that provide 15 or 20-amp overcurrent protection. Portable GFCIs should be used on a temporary basis when needed.

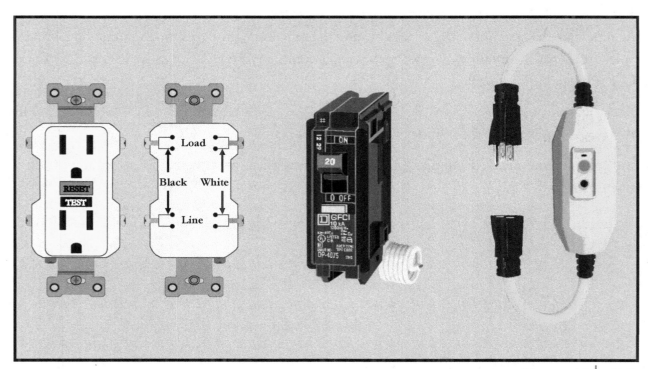

11.3 Areas Requiring GFCI

<u>In residential applications, GFCI receptacles are required to be installed in the following locations and appliances:</u>

- All 15 or 20-amp receptacles that are placed in bathrooms including those used as part of bathroom light circuits.

- Dishwashers, washing machines, and dryers

- Laundry areas

- Outdoor receptacles

- Crawl space receptacles

- Unfinished basements

- Kitchen countertop areas

- Pool motors and pumps

- Any receptacle within 6" of any sink

- Any boathouse

<u>The following are exceptions to the installation of GFCIs</u>

Receptacles that are not easily accessible and are supplied through branch circuits which are solely used for ice-melting or pipeline heating equipment are allowed to be installed without GFCI protection according to the *NEC®*.

Receptacles for garbage disposals and refrigerators are not required to be GFCI protected if they are not placed on the countertop in the kitchen area.

Receptacles that are dedicated to fire alarms or burglar alarm systems do not require GFCI protection.

11.4 How AFCI's Work

Arc Fault Circuit Interrupters are designed to prevent electrical fires in situations where arc faults occur. An arc fault is an unforeseen arcing condition that can happen in electrical circuits which can create a large amount of heat. Over time, this high heat can cause surrounding devices, wires, or framing to catch fire. Arc fault currents are usually caused by damaged wiring or device cords. This can happen during installation or over time.

AFCI devices contain sensitive components that constantly monitor if there are any dangerous arc faults. If the arc fault reaches a certain threshold of energy, then the AFCI will trigger to cut the flow of electricity and prevent any more heat from building up.

Unlike GFCI receptacles, AFCI receptacles must have metal conduit protection running from the circuit breaker to the first AFCI receptacle. Any receptacle found downstream after the first AFCI receptacle can be connected without conduit and will be AFCI protected.

Nail or Screw Penetration

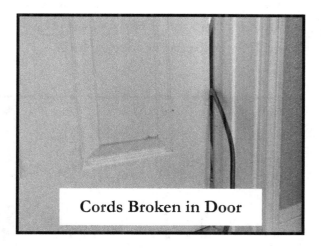

Cords Broken in Door

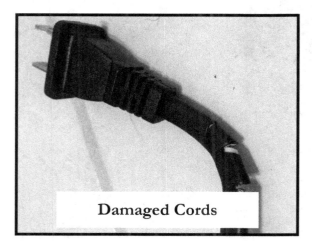

Damaged Cords

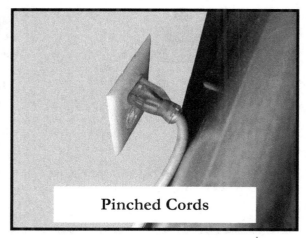

Pinched Cords

11.5 Types of AFCIs

There are four main types of AFCIs used in homes. The AFCI receptacle, AFCI circuit breaker, combination AFCI breaker, and the portable AFCI.

AFCI Receptacle - AFCI receptacles are installed in place of normal duplex receptacles. Similar to GFCI receptacles, AFCI receptacles can protect outlets downstream. The wires should be connected to the circuit breaker through metal conduit.

AFCI Circuit Breaker - Breakers installed in the electrical panel which protect against arc fault events. AFCI circuit breakers can only detect line to neutral or line to ground arc faults. These breakers can be either 15A or 20A at 120V for single-phase wiring.

Combination AFCI - Combination AFCI breakers combine the function of branch feeder AFCIs and receptacle AFCIs which can provide AFCI protection for devices plugged into protected receptacles.

Portable AFCI - Provide convenient AFCI protection when needed on the jobsite or for mobile applications. They can be used with other power strips or extension cords. They offer 15A or 20A protection at 120V.

11.6 Areas Requiring AFCI

With the 2017 update of the National Electric Code®, all receptacles that are installed in dwelling areas of residential structures must be AFCI protected. These areas exclude bathrooms, garages, and outside areas which are usually protected by GFCI. The *NEC®* also requires that replacement installations of circuit breakers or receptacles in these areas must be switched for AFCI protected devices.

The following are areas where AFCI protection is required:

- Kitchen
- Family Room
- Dining Room
- Living Room
- Bedroom
- Sunroom
- Library
- Den
- Office
- Hallways
- Closets

These areas are protected by AFCI devices which can detect if parallel or series arc faults occur. Parallel arc faults occur between two conductors while series arc faults occur between two sides of the same conductor. AFCI receptacles or circuit breakers protecting these areas can be tested and reset with buttons on these devices.

Parallel   Series

11.7 Installing GFCI or AFCI

GFCI or AFCI Receptacle Installation

The installation of GFCI or AFCI outlets can be done by following the manufacturer's guidelines for each GFCI or AFCI receptacle. In most cases, GFCI or AFCI duplex receptacles are normally installed similar to regular receptacles in standard device boxes.

You should follow normal procedures for the wiring of GFCI or AFCI receptacles. There are two line terminals to connect the incoming voltage into the receptacle. **There is also one load terminal used as the outgoing connection when used to connect to other downstream receptacles.** These must be wired properly or else the GFCI will not provide its intended protection.

The diagram below shows how a GFCI or AFCI receptacle can provide protection for other receptacles downstream. The load terminal wires from the GFCI or AFCI receptacle are connected to the terminals on the duplex receptacle to provide protection.

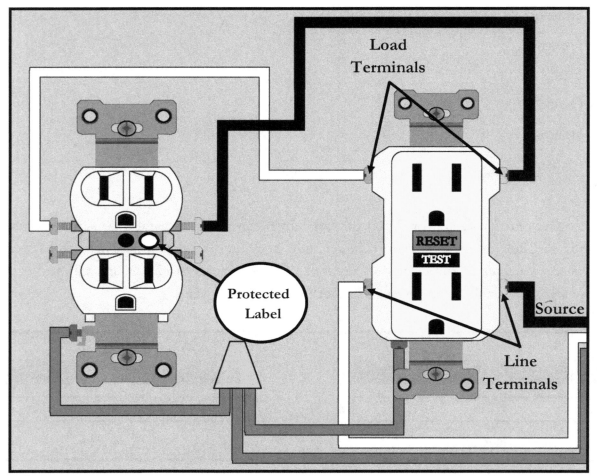

GFCI or AFCI Circuit Breaker Installation

1. After you have installed the GFCI or AFCI breaker in the proper slot in the electrical panel, ensure that the breaker is switched to the off position.

2. Connect the hot (black/red) wire to the hot screw terminal on the GFCI or AFCI circuit breaker.

3. Connect the neutral (white) wire to the neutral screw terminal on the GFCI or AFCI circuit breaker.

4. Connect the GFCI circuit breaker's coiled white wire to the neutral bus bar in the service panel.

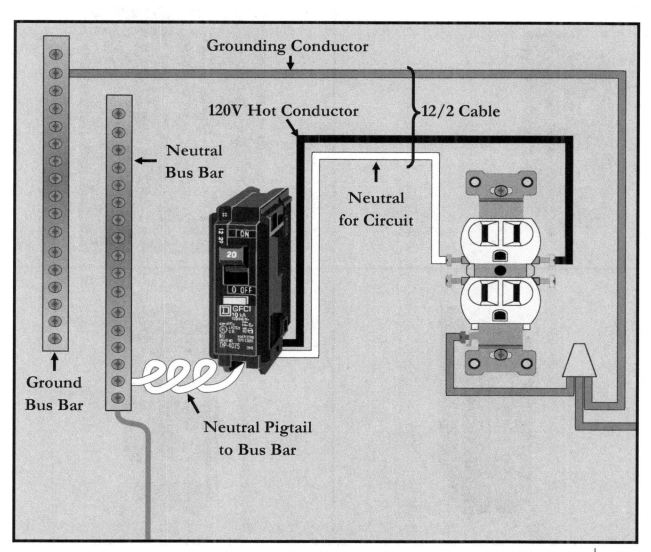

11.8 Testing GFCI or AFCI

GFCI or AFCI Circuit Breaker Testing

1. Make sure the circuit breaker is supplying power to a device.

2. Open the service panel and locate the breaker that will be tested.

3. While the breaker is on, press the test button.

4. If the breaker switches to the tripped position then the device is working properly.

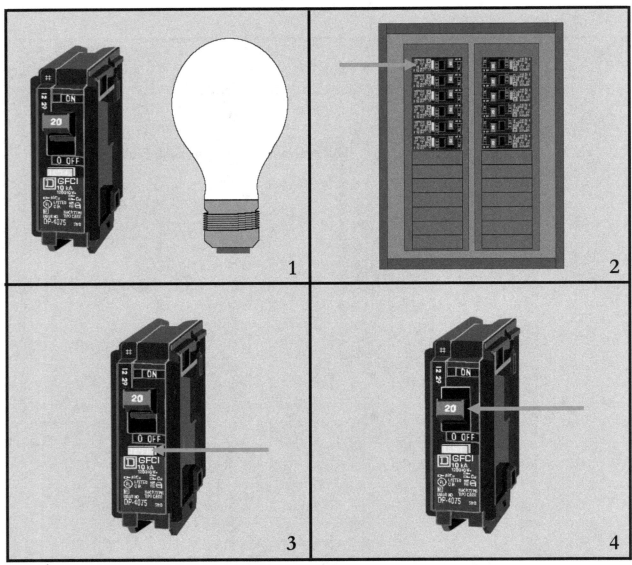

GFCI or AFCI Receptacle Testing

Proper installation of GFCI or AFCI receptacles will provide protection from ground faults and arc faults. Improper wiring of the receptacle prevents this protection and can be extremely dangerous. To prevent this, the receptacle comes with a testing mechanism. **If you accidentally connect the line wires to the load terminals then the testing mechanism on the receptacle will not work properly.**

1. The GFCI or AFCI is already tripped when it is purchased from the manufacturer. This prevents the receptacle from being reset until it has been properly installed. Wire the receptacle and finish the installation.

2. Plug in a device to the receptacle and turn on the circuit breaker. Press the RESET button on the receptacle. If the indicator light on the receptacle is on and the device is still off or does not have power you have incorrectly wired the receptacle.

3. If the device plugged into the receptacle has power press the TEST button. If the test button does not trip the receptacle by turning the indicator light off and cutting power to the device, then it has been incorrectly wired. If it does cut power, then the receptacle has been wired correctly.

11.9 Chapter 11 Review

1. A GFCI can detect as little as _____ milliamps of imbalance.

2. GFCIs or AFCIs are in place to protect _____, not electrical systems.

3. Two types of GFCIs are _____ and _____ .

4. _____ GFCIs can be utilized while using power tools on the jobsite.

5. Two areas that require GFCIs are _____ and _____ .

6. AFCIs detect _____ .

7. One way that arcing of a circuit can occur is _____ .

8. Two types of AFCIs are _____ and _____

9. Two areas that require AFCIs are _____ and _____ .

10. The two types of arc faults are _____ and _____ .

CHAPTER 11 NOTES:

Index

CPSIA information can be obtained
at www.ICGtesting.com
Printed in the USA
FSHW04n0317070318
45215FS